H.Ju. Ismajlova

Achievements of Azerbaijani scientists in the development of catalytic cracking

H.Ju. Ismajlova

Achievements of Azerbaijani scientists in the development of catalytic cracking

ScienciaScripts

Imprint

Any brand names and product names mentioned in this book are subject to trademark, brand or patent protection and are trademarks or registered trademarks of their respective holders. The use of brand names, product names, common names, trade names, product descriptions etc. even without a particular marking in this work is in no way to be construed to mean that such names may be regarded as unrestricted in respect of trademark and brand protection legislation and could thus be used by anyone.

Cover image: www.ingimage.com

This book is a translation from the original published under ISBN 978-620-7-80507-5.

Publisher:
Sciencia Scripts
is a trademark of
Dodo Books Indian Ocean Ltd. and OmniScriptum S.R.L publishing group

120 High Road, East Finchley, London, N2 9ED, United Kingdom
Str. Armeneasca 28/1, office 1, Chisinau MD-2012, Republic of Moldova, Europe
Printed at: see last page
ISBN: 978-620-7-85141-6

ACHIEVEMENTS OF AZERBAIJANI SCIENTISTS IN THE DEVELOPMENT OF CATALYTIC CRACKING

Ismailova H.Yu.

CONTENT

Keywords

Catalytic cracking, oil refining industry , achievements, development of science, plant modernization, processing depth, gasoline fraction, yield, octane number, increase

1. *ABOUT THE HISTORY OF EDUCATION AND DEVELOPMENT OF CHEMISTRY AND PETROCHEMICAL SCIENCE IN AZERBAIJAN*

Today in the national economy of Azerbaijan there is no industry where chemical industry products are not used.

In the era of the heyday of scientific and technological progress, chemical science and the petrochemical industry are rapidly developing and achieving new successes.

In Azerbaijan, the idea of chemistry is very ancient. Both in some eastern countries and in Azerbaijan, the creation of chemistry as a science began with ore mining and metal processing. Metallurgy, being the most important area of industry, is in turn associated with chemical transformations.

In order to determine the history of chemistry as a science, archaeological excavations have been carried out in Azerbaijan in recent years and, having studied a large number of material and cultural samples dating back to the Neolithic, Bronze, Iron and Middle Ages, scientists have come to the conclusion that the use of metal dates back to the end of the 4th millennium, the beginning of the 3rd millennium. According to archaeologists in ancient Azerbaijan, the first acquaintance of the population with metal begins with copper ore / 6-18 /.

In the Lesser Caucasus Mountains located on the territory of Azerbaijan, there are deposits of copper ore; methods for obtaining copper from them were known to our ancestors. It was established that back in the 2nd millennium , Azerbaijan was characterized by a higher bronze culture, and before Christmas in the 1st millennium, the Iron Age began. The most important iron ore deposits were located in Karabakh, especially near the Bayan village of Ganja district. This information has come to us in the scientific works of the historian Gamdullah Gazvini.

In the state of Manna, located on the territory of ancient Azerbaijan, copper, gold, silver and other metals were mined. Albanian historian M. Kalankaituklu wrote: "This country has all possible mineral resources. Gold, silver, copper and pure clay are mined in the mountains"/6/.

In the Middle Ages in Azerbaijan, medical science had a positive influence on the development of applied chemistry. In those days, along with herbal medicines, they also used medicinal substances prepared on the basis of oil, mercury and sulfur.

Prominent figures of science of the Middle Ages, the famous physician and scientist chemist Abu Ali Ibn Sina, and the scientific ideas of the scientist <u>Nasreddin</u> Tusi, who possessed encyclopedic knowledge, had a great contribution to the development of chemical science in Azerbaijan; founder of medical and applied chemistry Omar Osmanoglu.
In the 7th century in Azerbaijan, Omar Osmanoglu was a philosopher, mathematician, chemist and doctor.

Oil plays a major role in the education and development of chemical science and industry - oil production has an ancient history as a valuable natural resource in Azerbaijan. Azerbaijan is known as "the land of black gold and the land of fire". The first information about the production of Azerbaijani oil and its use for various purposes is found in the works of the Arab scientist and traveler Masudi, who lived in the 10th century. Even in ancient times, oil was mainly used as fuel and medicine for treatment / 6 /.

Over the past 100 years, Azerbaijan, located in an advantageous geographical position, in the development of rich oil and gas reserves, in the implementation of huge projects for exporting oil to the world market, has today turned into a geopolitical center for the intersection of interests of the United States, European and Asian countries. Thus, the oil industry of Azerbaijan constitutes an important element of its economic and political life. For the first time in the world (11 years before the United States), in Azerbaijan in 1848, Russian engineer Semenov drilled an oil well using a technical method. After this, in 1863, they began to drill an oil well mechanically on the island of Pirallahi. In 1869, in Balakhany, master Allahyar, when drilling a well mechanically, at a depth of 10-15 meters, a strong roar was heard, frightened by this, the master and workers closed the well. Three years later, in 1871, a well was drilled in Balakhany using a technical method, which produced 45,000 poods (700 tons) of oil per day /5/.

At the end of the 18th century, in the middle of the 11th century, the German Baron Thomas laid the foundation for photogenic (light oil) oil production in Azerbaijan.

For the first time in 1863, Baku technician Javad Melikov began producing kerosene (white oil). Since the 60s of the 11th century , oil refining tanks were built in Baku for industrial production. In Baku, already at the end of the 70s of the 11th century, individuals and firms had about two thousand oil refineries. It should be noted that in the 11th century the Russian Empire extracted oil in the villages of the Absheron Peninsula. It was this fact that laid the foundation for the development of the Russian oil industry /18/.

In 1872, 5 million tons of oil were produced in the villages of Baku, and in 1900, over 661 million tons of oil, which is 95% of all oil produced in Russia and more than half of the oil produced throughout the world. It was the production of kerosene (white oil) in 1873 that reached 1,500 tons. In the same year, Baku technician A. Tebrizov developed a process for continuous oil production. In 1875, kerosene production in Baku reached 847.7 thousand poods. In 1901, this figure reached 2,500,000 poods/5/.

Large oil production in Azerbaijan has attracted the attention and interest of foreign countries.

The great countries of the world competed with each other over Azerbaijani oil. Large companies flocked to Baku, such as: " StandartOyl ", American D. Rockefeller , " RoyalDeycaSall " under the leadership of G. Detelnitz from England, "Khazar-Black Sea" and "Mazut" of the Jew Rothschild from France, the Swiss syndicate " Nobel Brothers". In the development of the oil industry, the Nobel Brothers company in Azerbaijan had such significant role as the StandardOyl company for the USA. They organized a large-scale oil-industrial company "Partnership ", the founders of this company were the brothers Ludwig, Robert, Alfred Nobel and their friend Peter Bilderling / 5-12 /.

At the end of the 11th century in the Russian Empire, the head of the company, Ludwig Nobel, received the name "oil king" /18/. In Baku, owning an oil refinery, the Nobel brothers transported kerosene by sea through the Caspian Sea to the industrial centers of Russia and exported it abroad. In Russia, this company's kerosene sales amounted to 50.1%. Over the years, this figure reached 80.3%.

In the 11th century in Azerbaijan, the oil production and oil refining industry developed in a single economic system; its fate was inextricably linked with the successes and failures of the technological process of that time. In terms of technologies, scientific research, and equipment created, Azerbaijan does not lag behind world standards, but is also ahead in many ways /5/.

2. *THE ROLE OF OUR SCIENTISTS IN THE HISTORICAL DEVELOPMENT OF CHEMISTRY AND PETROCHEMICAL SCIENCE IN AZERBAIJAN*

The development of the petrochemical and oil refining industry in Azerbaijan was inextricably linked with the research work of scientists working in this field. The history of the development of science shows that the work of a researcher is not at all easy. It combines many years of scientific observations, scientific and practical research. The history of science reflects the greatness of work, nobility and deep scientific knowledge of scientists in materials about the bright, rich creativity of scientists.

Today, special attention should be paid to studying the historical development of chemical science and the petrochemical industry of Azerbaijan. It is our duty to inform the whole world and our future descendants about the noble work of scientists and scientific figures working in this field, their bright memory in the history of science.

In the development of chemistry and petrochemical science, scientists - chemists - Yu.G. Mamedaliev, M.F. Nagiev, V.S. Aliev, G.G. Gashimov, D.A. Guseinov, A.D. Lemberansky, R S.Sh.Kuliev, M.I.Rustamov, S.D.Mekhtiev, B.A.Dadashov, I.M.Orujova, B.K.Zeynalov, S.A.Sultanov, V.M.Abbasov, A.G .Azizov, N.M. Seidov, M.A. Mamedyarov, V.M. Akhmedov, F.I. Samedova, T.N. Shakhtakhtinsky, E.T. Suleymanova, R.A. Alieva, Sh.S. Vezirov, S.M. Aliev, G.T. Farkhadova, A.M. Seidrzaeva, A.J. Guseinova , E.G. Ismailov and others, their fundamental research works will fit into the memory of history, and today the most significant and main problem of our time is writing the history of natural and technical sciences in the range of the humanities. Valuable research in the field of oil refining, discoveries and inventions of Azerbaijani scientists, as well as books about their life's creativity can take a significant place in the development of chemical science and the petrochemical industry of Azerbaijan /22/.

In 1960, the book "Development of Science in Azerbaijan" published by academician Yu.G. Mamedaliyev about the history of science still does not lose its relevance and significance today, and is considered a valuable work on the history of science /17/.

In this book, Academician Yu.G. Mamedaliev pointed out that the development of chemistry as a science begins in the 12th century . He wrote: "The Azerbaijani people have raised a galaxy of scientists , thinkers, and poets. They enriched world culture and science with wonderful creations of human genius."

He further wrote: "... the great thinkers of the 10th century Abulhasan Bakhmanyar ibn Marzban and Khatib Tabrizi, the outstanding astronomer of the 12th century Farieddin Ali ibn Abdul Karim Shirvani and the famous engineer, scientist Emireddin Masud Nakhchievani, the physician and philosopher of the 12th century Efzaladdin Abulmelik Khunji left the most valuable scientific works that have survived to this day. » /17/.

In his book, Academician Yu.G. Mamedaliyev gave detailed information about the scientific activities of Azerbaijani chemists Movsum-bek Khanlarov and Sadikh Huseynov /17/.

Years pass, and chemical science in Azerbaijan is developing rapidly, the number of scientists working in this field is increasing. Currently, the role of national personnel in the development of chemical science is satisfactory.

Azerbaijan, being a country of oil, creates favorable conditions for the development of petroleum chemistry and oil refining. Today, as before, among the Azerbaijani scientific chemists we can proudly name the names - A.M. Maharramov, V.M. Abbasov, A.G. Azizov, N.M. Seidov, M.A. Mamedyarov, V.M. . Akhmedov, F.I. Samedova E.T. Suleymanova and others. It should be noted that the ongoing fundamental scientific research in the development of chemical science has found fame not only in Azerbaijan, but throughout the world.

Academician V.S., who has lived in Azerbaijan for many years. Gutyrya, together with other Azerbaijani scientists, conducted precise theoretical and practical research on various aluminosilicate catalysts for hydrocarbon conversion processes, and developed industrial methods for catalytic cracking on a "fluidized bed" catalyst /3/.

At the beginning of his career, Academician V.S. Gutyrya achieved success in developing a liquid-phase cracking technique. Then, in the mid-30s of the twentieth century, (together with M.A. Dalin), he developed a new modern technological process for producing ethyl alcohol from petroleum gases and actively participated in its industrial implementation.

The outstanding scientist V.S. Gutyrya worked for many years and led one of the large laboratories of the Azerbaijan Scientific Research Institute for Oil Refining named after V. Kuibyshev (current Institute of Petrochemical Processes named after Academician Yu.G. Mamedaliev of ANAS).

Working in Azerbaijan, academician V.S. Gutyrya wrote and published scientific works and monographs "Cracking and reforming of petroleum products using an aluminosilicate catalyst" (1944), "Catalytic purification of thermal reforming distillates" (1946), "Research in the field of catalytic cracking of petroleum feedstock using fluidized bed technology "(1962) /3-6/.

Even during World War II , the great task of supplying the front with fuel fell upon the lot of Azerbaijani oil workers.

During the war, Baku oil powered planes and tanks, resulting in an end to the war. It must be said with pride that during World War II , oil workers and scientists of Azerbaijan , using a number of inventions in the oil refining industry, supplied the Soviet Army with fuel and lubricating oils. Azerbaijani scientists and engineers have developed a technology for producing high-octane gasoline. In those years, 75 million tons of oil, 22 million tons of gasoline and other petroleum products were sent to the Union Fund /6-12/.

In 1941-1945, scientists of the current Institute of Petrochemical Processes named after Acad. SOUTH. Mamedaliyev ANAS created many inventions and discoveries.

During the Great Patriotic War, academician Yu.G. Mamedaliev's successes in the production of explosive substances and high-octane gasoline and aircraft fuel played a huge role in the victory over fascism.

During the war years, a group of scientists conducting active research was led by Ali Kuliyev. To obtain a number of explosives, they developed appropriate processes and organized the production of large-scale explosive ampoules in bottles /6/.

In the 60-70s of the twentieth century, under the leadership of academicians V.S. Aliyev and M.I. Rustamov, along with the improvement of catalytic cracking processes in a "fluidized bed" with a finely dispersed catalyst, intensive methods of catalytic cracking of petroleum feedstock were developed.

Academicians V.S. Aliev and M.I. Rustamov developed various modifications of reactor systems with descending through and semi-through flows, which made it possible to implement the process of two-stage catalytic cracking. The authors thoroughly investigated the development of a two-stage semi-through flow catalytic cracking process with gas-dynamic technological and thermal properties and introduced the process into industry. Together with co-authors, for the first time in the world, catalytic dehydrogenation processes were developed to obtain monoolefins by dehydrogenation of C_4-C_5 paraffinic hydrocarbons "in a fluidized bed" on a fine catalyst K-5 /22/.

The introduction of this process into industry contributed to a reduction in the production of divinyl from alcohol and, thus, when producing divinyl from synthetic alcohol, the cost of rubber decreased by 35% and the capital invested in building the plant decreased by 40-50%. In the process of this application, economic profit amounted to tens of millions of manats per year /2/.

This process was applied at many industrial synthetic rubber plants of the former Soviet Union - in Sumgait, Sterlitamak, Kuibyshev, Omsk, also in Romania.

Under the leadership of Academician V.S. Aliyev, valuable research was carried out and, on their basis, technological processes were created that contributed to the creation of a material base for the production of high-quality motor fuels. Academician V.S. Aliyev, by studying distillates of petroleum raw materials, oil residues and crude oil, contributed to increasing the reserves of raw materials.

Two processes developed by the scientist (two-stage catalytic cracking and an alunite processing method), licensed and protected by an author's certificate from the Committee on Patents and Discoveries of the USSR, were sold to the People's Republic of Bulgaria and the USA /23/.

A prominent scientist, academician M.I. Rustamov, one of the first researchers in mathematical modeling and optimization of oil refining processes.

He developed deterministic mathematical models of catalytic cracking processes of vacuum stripping, fuel oil, and oil, solved optimization problems under conditions of unsteady process flow, and compiled control schemes for these processes.

Under the direct supervision of Academician M.I. Rustamov, research work was carried out on the theory and practical implementation of catalytic processes with a finely dispersed catalyst; he laid the foundations for the development of a technological system for the dehydrogenation of butane into butylenes, the oxidation of o-xylene into phthalic anhydride, and progressive technological systems for catalytic cracking of various types of petroleum raw materials. The use of various modifications of reactor systems developed by him with rising through and semi-through flows made it possible to increase the specific productivity of the reaction apparatus by 8-10 times, significantly increasing the selectivity of the catalytic cracking process in comparison with the option of implementing it in systems with a "fluidized bed" of catalyst. In technological science, M.I. Rustamov was the first to introduce such concepts as "fluidity index", "critical fluid concentration of fine particles", "semi-through flow" /22-23/.

For many years, Academician M.I. Rustamov led research in the development of new effective technologies in oil refining and petrochemical processes that have found industrial implementation.

The processes that the scientist led were widely used not only at the oil refineries of our republic, but also in the cities of the former Union and at enterprises in Romania, Poland and Bulgaria /22-23/.

In the 70-80s of the twentieth century, in the scientific research of Academician M.I. Rustamov and Professor A.J. Guseinova, they found a solution to improve the quality of motor gasoline on a global scale. Thus, for the first time, AI-93 motor gasoline was obtained in the catalytic cracking process itself, and 1-B industrial-type catalytic cracking units were designed, they found application at the Azerneftyanajag production enterprise named after. G. Aliyeva /20/.

Academician M.I. Rustamov together with Doctor of Technical Sciences A.M.Seidrzaeva introduced into industry oil fractions of catalytic cracking with the participation of activators and passivators, and developed complex processing schemes /23/.

Academician M.I. Rustamov, Prof., Doctor of Technical Sciences G.T. Farkhadova, G.A. Abadzade is responsible for the development of destructive hydroisomerization and destructive isomerization processes aimed at producing low molecular weight hydrocarbons, which are successfully operated in catalytic cracking complex units /22/.

The development of the destructive isomerization process opens up new aspects of the development of the catalytic cracking process. At the same time, in contrast to the traditional purpose of the catalytic cracking process, the intended purpose of which is the production of motor gasoline, as the main process of the petrochemical process, the main purpose of which is the production of low molecular weight isoparaffin and isoolefin hydrocarbons. These processes are called destructive isomerization and hydroisomerization due to the fact that it is not the original hydrocarbon that undergoes isomerization, but the actual products of its cleavage (destruction) /7-8/.

The process of destructive hydroisomerization, developed with the participation of a newly synthesized nickel aluminosilicate catalyst, has received patents from leading capital countries - the USA, England, France, Canada, Germany, Japan /7,8,22/.

At the end of the twentieth century, ongoing research in the field of catalysis established the relationship between the surface properties of celite-containing catalysts, coking and the influence of these factors on the activity and selectivity of the catalytic cracking process. A distinctive feature of the research was the study of these patterns in the first seconds of contact of the raw material with the catalyst.

The results obtained made it possible to model the catalyst according to the height of the reactor and to regulate the surface properties of the catalyst by its coking. This became the scientific basis for creating a technology for processing unstable gasoline, and was confirmed during the processing of pyrocondensate at the G-43-107 catalytic cracking complex /7,8,23/.

Our scientists, together with leading scientists of the republic, have developed and proposed a new technological complex for oil refining with a main catalytic cracking unit, which allows for non-stop oil refining to produce promising fuels and raw materials for petrochemicals with a fewer number of technological stages. It was first established that in ionic-liquid catalysts, as a result of oligomerization of C_6-C_{10}- olefins, hybrid parafin-naphthenic synthetic oils with a freezing point (-55 °C-35 °C), viscosity index (VI 120) and temperature combustion (204-250 °C) /7-8/.

This work, completed within the framework of economic cooperation between Azerbaijan and Ukraine, was implemented for the first time in the CIS countries.

In addition, for the first time in the presence of CO_2, with the introduction of the supercritical extraction process, oil is being purified from asphalt-resinous compounds and metals, oil and its heavy residues, and environmentally friendly, harmless technologies have been developed.

Using new technologies, purified oil and hydrocracking of heavy oil residues, the production of various petroleum products and the use of catalysts in these processes give effective results. In order to make it possible to sell petroleum products (fuels and lubricants) abroad and to improve quantity and quality indicators, extensive marketing research is carried out. It should be taken into account that the profit received from the sale of petroleum products and petrochemicals far exceeds the profit from the sale of crude oil /22/.

Currently, research work is being carried out to meet the environmental requirements of products of the 21st century.

Today we can proudly note that the Baku Petrochemical School presented fundamental scientific discoveries to world chemical science.

For a long time to come they will enrich the world chemical and petrochemical science with their brilliant successes.

3. *MAIN STAGES IN THE DEVELOPMENT OF WORK IN THE FIELD OF CATALYTIC CRACKING*

The main stages in the development of catalytic cracking, a list of which is presented in the works, are:

- transition from catalytic cracking in a stationary catalyst bed to cracking in a fluidized bed of a microspherical catalyst;

- development of a zeolite-containing catalyst;

- hydrotreating of raw materials;

- cracking in a riser reactor with an upward flow of microspherical catalyst;

- regeneration of the catalyst with complete combustion of CO into CO$_2$ in the regenerator.

The most important achievements in the field of catalytic cracking are inextricably linked with the discovery of the catalytic activity of zeolites and the creation of zeolite-containing catalysts based on them, which led to a sharp increase in the yield of the high-octane gasoline component. Currently, in our country, more than 80% of catalytic cracking units operate with a zeolite-containing catalyst or its mixture with an amorphous aluminosilicate catalyst.

The introduction of highly active and selective zeolite-containing catalysts revealed the need for cracking in riser reactors with an upward flow of microspherical catalyst, as a result of which maximum use of catalyst activity was achieved with minimal secondary cracking.

Currently, several options have been developed for the reconstruction of reactor-regenerator units of catalytic cracking units, aimed at increasing the productivity of raw materials and ensuring efficient cracking of raw materials on a microspherical zeolite-containing catalyst. Thus, in the GrozNII and INHP of the Academy of Sciences of the AzSSR, a technological scheme has been developed that provides for the cracking of recirculated raw materials in two vertical lift reactors ending in zones of a forced fluidized bed /24/.

According to the two-stage cracking scheme with a combined reactor developed at VNIINP, the cracking of raw materials is carried out in a riser reactor, and then in a step-countercurrent zone in contact with fresh catalyst supplied to the upper section.

The Institute of Chemical Industry of the Academy of Sciences of the AzSSR proposed cracking of the feedstock in recirculated gasoline fractions in two separate reactors. After regeneration, the catalyst enters both reactors for contact with raw materials and recycle, and cracking products are discharged into separate distillation columns for separation.

$^{\circ}C$ fraction as a recirculate, it is possible to obtain about 46% by weight of gasoline that meets the requirements when presented with AI-93/20-21/ gasoline.

A significant improvement in the performance of the catalytic cracking process is achieved by special hydrogenation preparation of raw materials and circulating gas oils (recycle). At the same time, along with an increase in the degree of conversion of raw materials and gasoline yield, the sulfur content in gasoline and sulfur oxides in the exhaust gases sharply decrease.

...Grozgiproneftekhim, based on research by GrozNII, VNIIP, INHP AN AzSSR, has developed a scheme of a combined installation, including units for hydrotreating vacuum distillate, catalytic cracking, rectification of cracking products and gas fractionation.

In Fig. Table 1 shows the stages expected based on research for a gradual increase in gasoline yield with a constant coke yield of 6% wt.

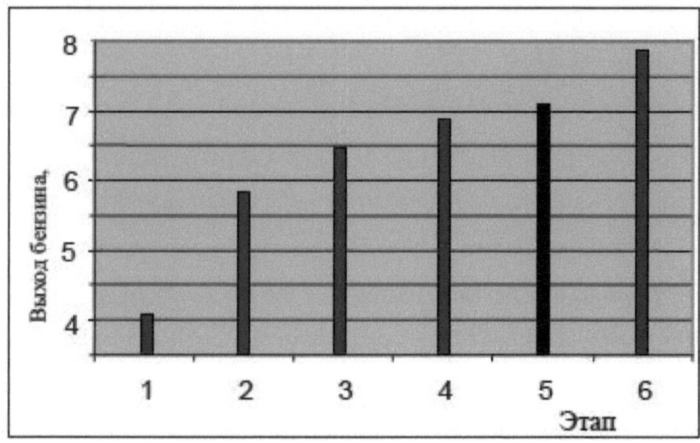

Fig.1. 1- amorphous catalyst , 2- zeolite-containing catalyst ; ; 3 - straight-run elevator reactor ; 4 - hydrotreated product ; 5 - highly hydrotreated product; 6 - hydrotreated recycled gas oil.

As can be seen from the figure, the maximum gasoline yield can reach 77% vol. or ~62% wt. It is assumed that the theoretically achievable - maximum gasoline yield during catalytic cracking is ~85% vol.

The practical importance of the research results of recent years (as a result of the use of improved catalysts, changes in cracking and regeneration conditions, modernization of equipment) is obvious, because in existing catalytic cracking units the gasoline yield reaches only 48-50% by weight.

The issue of improving catalysts is covered in detail below. It should be noted here that improving the performance of catalytic cracking also contributes to improving the design of the reactor-regenerator unit. For example, according to data, multipoint input of raw materials into the riser reactor significantly improves the distribution of the catalyst and the temperature profile in the riser reactor, and also promotes better contact of the catalyst with the feedstock. As a result, the degree of conversion and the yield of gasoline respectively increase by 2.5 and 2.0% vol. while simultaneously improving the octane characteristics of the latter.

Another important trend in the development of catalytic cracking is the heavier weight of the processed raw materials - the involvement of fuel oil and oil in the processing. It should be noted that due to the fact that one of the reasons for the limited use of cracking oils and fuel oils is the high content of metal impurities in them, the improvement of catalyst demetallization processes becomes a necessary stage in the development of cracking of heavy raw materials.

The catalytic cracking process is highly flexible and, depending on demand, can be used to either produce maximum quantities of gasoline or middle distillates, or to produce petrochemical or alkylate feedstocks.

Considering the intensive development of scientific and technical work on the creation of artificial and synthetic fuels, the development of a petrochemical version of the catalytic cracking process is considered as new aspects, as horizons for catalytic cracking. In this direction, in 1973, at the Institute of Chemical Industry of the Academy of Sciences of the AzSSR, work began on the development of a petrochemical version of catalytic cracking. As a result of studying the influence of various parameters on the process (temperature, mass feed rate of raw materials, molecular weight and composition of the feedstock), a process called destructive isomerization was developed. The process is aimed at the production of low-molecular hydrocarbons of isostructure and is carried out in reactors with an upward flow of zeolite-containing catalyst Tseokar-2. The optimal raw material for the process is petroleum distillates with an average molecular weight of 140-180 and containing no more than 16-18% aromatic hydrocarbons. As one of the types of raw materials, coking phlegm (fr. 160-260 ° C) is successfully involved in the process.

Compared to processes of similar purpose (hydrocracking, pyrolysis, isomerization), the process of destructive isomerization can significantly expand the raw material resources for the production of low-molecular hydrocarbons of isostructure; it is simple in technological design, does not require pressure and the presence of hydrogen. In addition, in contrast to the processes of hydrocracking and hydroisomerization, in the process of destructive isomerization, along with saturated hydrocarbons of isostructure, isoolefinic hydrocarbons are also obtained. This is very important if we take into account the further processing of the resulting intermediates for the synthesis of monomers.

Maximizing the yield of isoolefin hydrocarbons will become possible either by minimizing the contact time of the raw material with the catalyst, or by introducing new zeolite-containing catalysts. /24/

3.1. IMPROVEMENT OF CATALYTIC CRACKING CATALYSTS BY AZERBAIJAN SCIENTISTS

As noted above, the creation of highly efficient catalysts with a set of properties that meet the requirements of modern technology is of paramount importance in the development of new catalytic processes.

Today, the main ones among the new catalytic systems are zeolite-containing catalysts, which have replaced amorphous aluminosilicates in almost all catalytic cracking units, which has made it possible to significantly increase the depth of conversion of raw materials while simultaneously increasing the selectivity in the formation of gasoline, as the main target product of the process.

The first commercial zeolite-containing catalysts contained faujasite type zeolite (molecular sieves "A"). Currently, 80% of industrial zeolite-containing cracking catalysts are made on the basis of "U" type zeolite in a rare earth exchange form, introduced into a synthetic amorphous aluminosilicate matrix or its mixture with kaolin clay.

The largest manufacturers of zeolite-containing catalysts in the United States are Davison, American Cyanamid, Philtrop Corporation, Nalio Corporation, Goodry, and Mobil Oil; in Europe - the company "Ketjen" (Holland).

Among the American zeolite catalysts for moving bed cracking, the series of zeolite-containing catalysts of the Lurabed brand should be noted. Using the Lyurabed-5 catalyst, an increase in gasoline yield by 12% vol. was achieved compared to amorphous aluminosilicate. raw materials with less gas and coke formation. However, this catalyst had a low bulk density and abrasion resistance. When producing the next catalyst, "Lurabed-6," for cracking in a moving bed, finely dispersed aluminum oxide (-form) was introduced into the amorphous almosilicate matrix along with zeolites α, which ensured an increase in its bulk density by 7%. It is also worth noting the high-alumina catalysts "Lyurabed-7" and "Lyurabed-8" based on "U" type zeolite, the high strength of which is also achieved by introducing highly dispersed Al_2O_3 as a sealant during the coprecipitation αprocess .

Among the zeolite-containing catalysts for fluidized bed cracking processes, one can note the D-9 microspherical catalyst containing decationized zeolite (REE), which has good selectivity and resistance to introduced steam and high temperatures.

Microspherical REE catalysts XZ-15, XZ-25, XZ-36, XZ-40 were obtained by spray drying of a mixed, crystalline amorphous gel and are characterized by a high zeolite content. The XZ series catalysts are characterized by a sharply reduced sodium oxide content, which is very important due to the increase in zeolite content, which can significantly increase the activity and selectivity of the catalysts.

The high-alumina zeolite-containing catalyst H Z -1 based on cheap natural kaolin raw materials has high stability and high mechanical strength (due to the introduction of a filler - aluminum oxide), and in terms of activity it is close to synthetic zeolite-containing catalysts of the Dorabed series.

Zeolite-containing catalysts HFZ-20 and HFZ-23 , containing 52-59% by weight of aluminum oxide, are characterized by high activity, increased density (0.86-0.90 kg/l) and reduced sodium oxide content.

Among American industrial cracking catalysts, I would also like to mention the zeolite-containing catalyst CO-1 based on a magnesium silicate matrix. It is noted that with the same general activities of catalysts based on magnesium silicate and semi-synthetic aluminosilicate matrices, the presence of magnesium silicate greatly affects the yield of cracking products.

The latest modifications of the catalysts burn carbon monoxide to dioxide in the regenerator (SS Z -22.44, RS Z -22.44, Koukat), bind sulfur compounds of oxide and sulfur dioxide in the regenerator (Rezidkat-20, Sokskat), increase the octane number of gasoline (Octaket) and are resistant to processing raw materials with a high content of heavy materials (GRZ - 1, Residecat-20, Residecat-30). According to the data, in the presence of the SS Z -22 catalyst, carbon monoxide is almost completely oxidized to CO_2 in the regenerator, and the content of residual coke on the catalyst itself is significantly reduced.

Special additives have been developed to bind sulfur oxides in the regenerator, which are included in zeolite-containing cracking catalysts and do not affect their activity and selectivity. The sulfur-containing compounds formed on the catalysts, when entering the reactor, decompose to form the initial additives of hydrogen sulfide , which is removed by ethanolamine purification.

Along with the introduction of promoters directly into zeolite-containing catalysts, abroad, in order to reduce the harmful effects of metals contained in raw materials, a liquid additive based on antimony compounds is introduced into the reactor, which, without dissolving in cracking products and without interacting with water vapor, is adsorbed on zeolite-containing catalysts , preventing them from being poisoned by heavy metals. The use of the additive simultaneously with an increase in the service life of the catalyst makes it possible to increase the yield of gasoline by 6% while reducing the yield of coke, hydrogen and dry gas. /4/

In recent years, the company "Ketjen" (Holland) has been producing catalysts M Z -3, M Z -3 S with a high content of zeolite type "U" of the rare earth form. In terms of its composition, the M Z -3 catalyst is distinguished by a high-alumina matrix and a low residual sodium oxide content. The catalyst M Z -3 S, due to its smaller pore volume, is more thermally stable than M Z -3./4/

The Soviet Union is also successfully working on the development of zeolite-containing cracking catalysts.

In 1968, the first zeolite-containing catalyst AMNTs-1 based on "X" type zeolite was produced and tested at VNIIIN, and in 1969, the Tseokar-1 catalyst was developed at GrozNII - the first of the rare earth modifications of the catalyst, the composition of which also used "type" zeolite X".

After the industrial development of the production of "U" type zeolite, on its basis, industrial bead zeolite-containing catalysts AMNTs-3 and Tseokar-2 were obtained, characterized in that the AMNTs-3 catalyst contains "U" type zeolite in ammonium form, and Tseokar -2 is synthesized on the basis of "U" type zeolite in a rare earth exchange form. As a matrix in both of these types of catalysts, a synthetic amorphous aluminosilicate containing 6-7 wt .% aluminum oxide /9,10,24/ is used.

The domestic zeolite-containing catalyst Tseokar-2 in such indicators as initial activity, stable activity, selectivity, mechanical strength, coke-forming ability is close to the similar foreign sample Durabed-8, second only to it in terms of bulk density (0.7 versus 0.8 g/cm ^{3}).

At present, samples of the Tseokar-2 catalyst have been prepared, in which an increase in the hired density is achieved by introducing fine fillers. In this case, catalysts were obtained having a bulk density of about 0.3 g/cm ^{3} , and the introduction of some fine fillers into a ball zeolite-containing catalyst in an amount of about 10% significantly increases the wear resistance of this catalyst.

On an industrial scale, the production of microspherical zeolite-containing cracking catalyst with rare earth elements KMTsR-2, which is a microspherical analogue of the Tseokar-2 catalyst, has been carried out. Increasing the stability of the KMCR-2 (KMCR-4) catalyst is achieved by increasing the content of the zeolite component to 20 wt. % and aluminum oxide to 18-20 wt.%.

For future production, the GrozNII has developed a technology for producing microspherical catalyst Microzeocar-5 for operation in cracking plants with lift reactors. The high thermal stability of this catalyst is achieved primarily by using a deeply substituted zeolite in a rare earth exchange form. In addition, during the synthesis of the Microzeocar-5 catalyst, specially developed highly stable matrices are used, containing an optimal amount of aluminum oxide (~ 40%) and having a higher stabilizing ability than the matrices of currently produced catalysts.

On the issue of the influence of the properties of zeolites on the activity of zeolite-containing catalysts in the cracking reaction of both individual hydrocarbons and individual oil fractions, extensive research has been carried out, as well as individual oil fractions, and extensive research has been carried out by Topchieva K.V. , Masagutovy R.M., Piguzovoy L.I., Rabo D., Dorogochinsky A.Z., Romanovsky B.V., Galich N.N., Rossolovskoy V.N./24/

As a result, it was found that the activity of zeolite-containing cracking catalysts depends on the structure of the zeolite, the amount and form of exchange cations, the order of introduction of cations into the exchange complex, as well as the processing conditions of the zeolite and the type of raw material being processed.

However, the catalytic properties of zeolite-containing catalysts depend not only on the properties of the zeolite and the ratio of active components, but also on the properties of the amorphous component. There is little information on this issue in the literature, and a review of the available material regarding the functions of the matrix in the composition of cracking zeolite-containing catalysts is of considerable interest.

3.2. IMPROVEMENT OF ZEOLITE-CONTAINING CATALYST CATALYTIC CRACKING PROCESS

As is known, at present, zeolite-containing cracking catalysts, which are two-component zeolite-matrix systems, where the zeolite component is uniformly distributed throughout the mass of the amorphous material, have found the greatest practical application.

Catalysis on zeolite-containing catalysts has a number of features that are not typical for pure zeolites. Here the role of the amorphous phase, the matrix, is manifested, which has a noticeable effect on the stability of the crystalline phase of the zeolite and, consequently, on the catalytic activity.

The literature provides various information regarding the role of the matrix in the composition of zeolite-containing cracking catalysts. Eastwood believes that the functions of the matrix are to mechanically envelop each crystal, remove heat from the zeolite, and give the catalyst the necessary diffusion properties. Niguzova L.I. The increased activity and stability of the catalyst, consisting of two crystalline phases of the same chemical nature, is explained by a synergistic effect, considering that the matrix is not an inert material, but plays a certain role as an active component.

The patent indicates increased cracking activity of a zeolite-containing catalyst prepared from kaolin compared to a catalyst prepared from aluminosilicate. Mage et al., as a result of studying the influence of the type of carrier on the properties of cracking catalysts, found that zeolite-containing catalysts obtained on the basis of various matrices exhibit different activity in terms of the yield of gasoline and unsaturated gases $C_3 + C_4$.

As is known, both components of a zeolite-containing catalyst—zeolite and aluminosilicate—are ion exchangers. This creates certain prerequisites for interphase transitions of exchangeable cations. Migration of cations between the zeolite component and the amorphous matrix becomes possible, which, in turn, leads to corresponding changes in the acidic and catalytic properties of zeolite-containing systems.

The migration of cations in the matrix-zeolite component system was studied experimentally by K.V. Topchieva. et al. by local X-ray spectral analysis. It has been shown that sodium cations migrate from the zeolite component into the matrix during thermosteam treatment. If the sodium content in the matrix is higher compared to the zeolite component, then the cations migrate in the opposite direction, i.e. During thermosteam treatment, a uniform distribution of sodium cations is observed between the crystalline and amorphous phases.

The calcium form of zeolite is characterized by less migration of calcium cations from the zeolite into the matrix, and in the case of lanthanum-substituted zeolite, migration of cations into the amorphous matrix during thermosteam treatment is practically absent, since lanthanum cations are much more bound in the zeolite structure.

These changes in the structure of the zeolite and the arrangement of exchangeable cations during thermosteam treatment affect the spectrum of acid sites, which, as is known, is essential for catalysis. Namely, high-temperature treatment reduces, first of all, the number of strong acid sites: acid sites of medium and weak strength change to a lesser extent. In general, during thermosteam treatment, a narrowing of the "acidity spectrum" of zeolites, as well as zeolite-containing cracking catalysts, is observed.

The matrix plays a great role in the formation of a stable catalyst. It has been established that pure zeolites are completely destroyed during heat treatment with water vapor, while in the matrix they retain a certain degree of crystallinity.

The effect of high-temperature treatment with water steam on the destruction of zeolite in the composition of an aluminosilicate catalyst was also studied by Ya.V. Mirsky. , which showed that zeolite processed outside the aluminosilicate carrier is completely destroyed already at 750 °C.

To a certain extent, the properties of zeolite-containing catalysts are also influenced by the chemical composition of the matrix. This is clearly observed when comparing zeolite-containing catalysts with magnesium silicate and aluminosilicate matrices. Thus, according to data at a conversion of West Texas gas oil equal to 77% vol., a zeolite-containing catalyst based on a magnesium silicate matrix makes it possible to produce 4.5 vol.% more gasoline than a zeolite-containing aluminosilicate catalyst with almost the same coke formation and 6.6% less vol. . gas outlet.

A comparison of the catalytic activity of magnesium silicate and aluminosilicate matrices that do not contain zeolite shows that the magnesium silicate system reduces the content of light hydrocarbons, increases the yield of gasoline, and significantly increases the yield and quality of light recirculating gas oil (with a boiling range of 220-340 $^\circ$C). This increase in the yield of light gas oil is caused by the suppression of secondary cracking reactions, which is confirmed by a decrease in the yield of light hydrocarbons, as well as aromatic and olefinic hydrocarbons. A decrease in the yield of the latter leads to a slight deterioration in the octane characteristics of gasoline.

The Institute of Chemical Industry of the Academy of Sciences of the AzSSR also conducts research on the influence of the nature of the matrix on the properties of zeolite-containing catalysts based on it. It has been shown that the nature of the magnesium silicate matrix and the activity of the zeolite-containing catalyst based on it have a significant influence on the thermal stability of the zeolite in its composition, on the pore-structural characteristics of the catalyst, on the acidic properties of its surface, as well as on the formation of the hydrocarbon composition of cracking products (gasoline and gas).

Machinskaya M.V. et al. conducted a study on the influence of the chemical composition of the matrix on the properties of zeolite-containing cracking catalysts by varying the SiO_2/Al_2O_3 ratio in the matrix, as well as by changing its cationic cover. It was discovered that the catalytic properties and physical structure of the aluminosilicate matrix change with a change in its chemical composition, on the one hand, and the catalytic properties of the corresponding zeolite-containing catalysts, on the other.

Zulfugarov Z.G. et al. investigated the influence of the nature of the matrix based on alkaline earth silicates on the activity of zeolite-containing catalysts and the composition of cracking products and showed the possibility of regulating the composition and yield of cracking products by changing the chemical composition of the matrix and zeolite-containing catalysts based on it.

The porous structure of the matrix can contribute to the maximum manifestation of the catalytic properties of zeolite-containing catalysts or, conversely, exert diffusion inhibition, ultimately leading to a decrease in activity and selectivity. In this regard, as noted above, one of the most important functions of the matrix is to ensure the transport of raw material molecules to the zeolite component, uniformly distributed throughout the entire mass of the amorphous material.

According to the data, the grain of a zeolite-containing catalyst is represented as a system of a bidisperse matrix with branching model pores and a zeolite component. It has been established that by reducing diffusion inhibition with an increase in the radius of large matrix pores, it is possible to achieve a significant increase in the productivity of zeolite-containing catalyst grains.

The influence of the effective pore radius of the matrix on the activity and selectivity of the industrial bead zeolite-containing catalyst Tseokar-2 during cracking of the light kerosene-gas oil fraction was also studied by K.V. Topchieva. and it has been shown that for the specified type of catalysts with a grain size of 3-5 mm, an increase in the zeolite content above 15-20 wt% does not lead to a significant increase in the activity of the catalyst, which is probably due to the inhibition of the transport of raw materials to the zeolite crystals in the pores of the matrix or in secondary porous structure of granular zeolite.

According to the data, the activity and selectivity of cracking catalysts with different volumes and radii of secondary pores are different, namely, the activity of zeolite-containing cracking catalysts decreases with a decrease in the volume of secondary pores.

These results highlight the importance of ensuring optimal secondary pore structure of zeolite-containing catalysts.

The work also indicates that zeolite-containing catalysts, characterized by different pore-structural characteristics, represent different catalytic systems. According to the data, the activity and selectivity of catalysts is determined by the nature of the micropore compounds that inevitably arise around zeolite crystals when it is introduced into an amorphous aluminosilicate gel.

Thus, the high activity of a mixed zeolite-containing catalyst is explained by the formation in the presence of a binder of a developed system of large transport pores, which sharply reduces the influence of slow diffusion stages on the process. Depending on the type of binder, a different porous structure of the mixed catalyst is created, the nature of which determines the speed of the process, as well as the rate of catalyst regeneration. Moreover, a number of studies indicate that by changing the nature of the matrix, targeted production of cracking products is possible. Thus, the work indicates that with the help of non-zeolite components it is possible to change some properties of zeolite-containing catalysts. Namely, the slightly higher yield of olefinic hydrocarbons on the zeolite-containing catalyst T S -4 compared to other zeolite contacts under identical conditions is explained by the specific properties of the matrix into which the zeolite component is introduced.

This is especially important in connection with the "olefin problem", since an increase in the yield of unsaturated hydrocarbons, as some authors suggest, by increasing the cracking temperature and feed rate, which reduce the intensification of the reactions of hydrogen redistribution between individual hydrocarbon molecules, causes a slight decrease in the selectivity of the formation of the main cracking product - gasoline.

Thus, taking into account that the catalytic and regeneration properties of zeolite-containing systems are largely determined by the nature of the matrix, the creation of binding components that contribute to the formation of wide-pore zeolite-containing catalysts with high activity and selectivity becomes important.

Currently existing industrial samples of zeolite-containing catalysts are prepared mainly on the basis of an aluminosilicate matrix. Although zeolite-containing magnesium silicate catalysts for cracking petroleum feedstocks can increase the yield of gasoline, their wide industrial distribution is limited due to their difficult regenerability. Attempts to improve the regeneration characteristics of magnesium silicate catalysts by increasing the pore diameter of the catalyst did not lead to positive results, since with an increase in the coke burning coefficient, the activity of the catalyst decreased. There is also information in the literature about the use of a magnesium silicate catalyst in a mixture with calcium oxide (1-18%) for cracking heavy hydrocarbons.

As a matrix of a zeolite-containing catalyst, this dissertation work uses a gel formed when natural waters of various origins are softened with sodium silicate. The use of the gel as a binder is dictated, first of all, by its chemical composition, presented mainly. Ca- and Mg - silicates containing small amounts of impurities of other metals. If we also take into account the unlimited resources of raw materials for producing the gel (natural waters), then the possibility of synthesizing zeolite-containing catalysts on its basis seems quite promising.

4. *DEVELOPMENT AND CURRENT STATUS OF THE CATALYTIC CRACKING PROCESS IN AZERBAIJAN*

Among the catalytic processes proposed for the processing of petroleum fractions, catalytic cracking occupies one of the first places in its importance.

In 1936, catalytic cracking was first used on an industrial scale and has since become the most important process in the oil refining industry of many countries.

The process of catalytic cracking of petroleum feedstock has currently undergone major changes in technological and equipment design.

The rapid development of catalytic cracking was facilitated by the successful combination in the process of wide possibilities for processing a wide variety of raw materials: from light distillate fractions to heavy vacuum gas oils and even residual oil fractions.

As is known, the first industrial catalytic cracking installation was the installation of the Goodry system with a stationary layer of aluminosilicate catalyst . Subsequently, more progressive catalytic cracking systems based on mobile granular and finely dispersed catalysts were developed abroad and in the USSR. /24/

This development of the process became possible thanks to the research conducted by V.S. Gutyrya, V.S. Aliyev, M.F. Nagiyev and many other scientists in our country, and the research of Egloff, Gaer, Kinsberg, Goode, Greensfeltser, Thomas and others spent abroad./25/

Existing industrial catalytic cracking systems can be divided into two groups:

1. with a stationary catalyst bed in batch reactors;
2. with a circulating catalyst in continuous reaction apparatuses.

In turn, the settings of the second group are divided into two subgroups:

1. with a moving granulated catalyst;

2. with a dust-like or microspherical catalyst that forms a fluidized bed in contact devices.

Each of the listed subgroups corresponds to a large number of installations, differing in hardware and technological design.

Without dwelling on the advantages and disadvantages of each of the systems, we will only note that due to a number of significant advantages of the catalytic cracking system with a fluidized bed of fine catalyst over the others, the latter has received the most widespread development in the oil refining industry.

The main advantages of this system include the following:

1. the main devices - the reactor and the regenerator - are structurally simpler and are essentially hollow vessels;

2. circulation of the catalyst in the system can be carried out with a constant flow of transporting vapors and gases in a wide range of speeds depending on the density of the gas catalyst flow;

3. intensive mixing of the masses of dusty catalyst and raw material vapors leads to equalization of the layer temperature and the elimination of local overheating;

4. it is possible to use various types of raw materials - from gasoline to heavy fractions of vacuum stripping;

5. As a result of the smaller particle size, intradiffusion inhibition, which limits the rate of catalyst regeneration, is significantly reduced.

The table shows the growth in catalytic cracking capacity of various systems in recent years in the United States.

It is interesting to note that the direct distillation capacity in the USA as of 1/1-1961 was 1341.0 thousand tons /day.

It follows that the total capacity of all catalytic cracking systems in the USA as of 1/1-1961 was 50.8% of the total straight-line capacity. Moreover, it is shown that the development of catalytic cracking in the USA is ahead of the growth of direct racing.

As for catalytic cracking with a fluidized bed of fine catalyst, then, as follows from Table 1, this process is the main one, and in 1963 it accounted for 79.9% of all raw materials processed in catalytic cracking units.

No.	Name of settings	Number of econluminous installations	Maximum power of 1 installation, t/day	Total installation force in 1960, t/day	Product yield							Octane number of gasoline
					From 5 and above,	Dry gas, volume %	C_3,	C_4, volume	Coke, all %	Lekium gasoil, volume	Heavy gas oil volume %	
1	Model 4	37	8400	103.0	43.8	3.5	7.1	10.0	7.0	17.3	18.9	92-99
2	Ortoflow (models A and B)	22	16000	85.0	-	-	-	-	-	-	-	-

| 3 | UOP . | 70 | 4950 | 89.0 | 45.2 | 2.2 | 7.9 | 12.3 | 7.0 | 28.0 | 12.0 | 98.5 |

Table

The industrial development of catalytic cracking in the USSR began with the construction of Goudry installations and the creation by 1964 of a domestic system with a moving granular catalyst. In subsequent years, a large number of installations were built, including successfully operating installations with dust catalysts of type 1-B and 1-A, developed by the Institute of Chemical Industry of the Academy of Sciences of the Azerbaijan SSR and Giproazneft./ 2/

Currently, fluidized bed installations are being built in the USSR, the power of which is twice as high as the capacity of the existing installation. Since the topic of our work is the research and development of technology for the two-stage catalytic cracking process, in which the first stage is carried out in a through flow, and the second in a fluidized bed, then in the first part of the literature review we will critically monitor the development of the catalytic cracking process with a fine catalyst, give assessment of the positive and negative aspects of different years of technological design of the process, and also consider the issue of the influence of the quality of raw materials (chemical and fractional composition) on the yield of target products during the cracking process.

4.1. DEVELOPMENT OF CATALYTIC CRACKING EQUIPMENT

The first industrial catalytic cracking unit with a circulating pulverized catalyst of the "model 1" type, the diagrams of which are shown in Fig. 1, was put into operation in 1942.

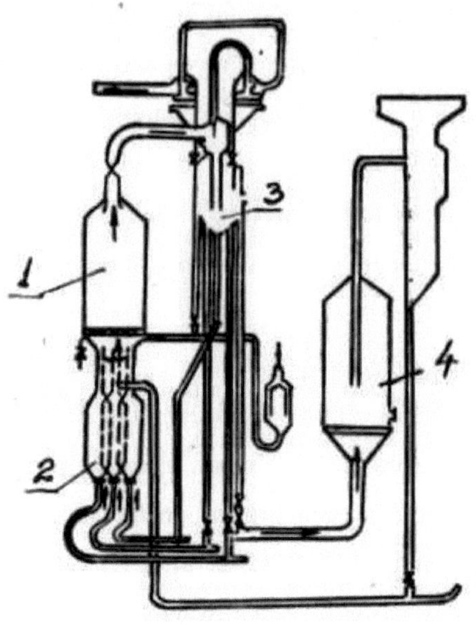

Fig. 1 Schematic and technical diagram of the installations. Model 1. 1-Regenerator, 2-refrigerator for regenerator, 3-hopper, 4-reactor

In this installation, the entire mass of circulating dusty catalyst passed through the reactor and regenerator from the bottom up and was removed entirely from the top of these devices in a mixture with gaseous reaction products. The most characteristic feature of Model 1 installations is the upper outlet of the circulating catalyst from the reaction apparatus. A stream of air or oil vapor is separated from the catalyst in cyclone separators.

These installations are interesting not only as the first industrial installations with a finely dispersed catalyst, but mainly because the reaction and regeneration here are carried out under continuous transport conditions at relatively high densities of the gas catalyst flow. Consequently, the principles of catalytic cracking in this installation are essentially the same as in comparison with contacting and reactions occurring in a fluidized bed.

In 1951, after the reconstruction of one of the model units, its capacity was increased to 7000 tons of raw materials per day. Judging by the literature, the installations of model 1 were not widely used and were supplanted due to the following main disadvantages inherent in them:

1. Significant length of catalyst wires and high hydraulic resistance of the catalyst circulation system.
2. The need to use large cyclone separators designed to capture the entire amount of circulating catalyst.
3. Increased abrasion of the catalyst and, consequently, its high consumption.
4. The bulkiness of the installation and its lack of operational flexibility. The height of this installation reached 70 meters.

The next step in the development of catalytic cracking technology was the appearance of units of the model 11 type, the diagram of which is presented in Fig. 2.

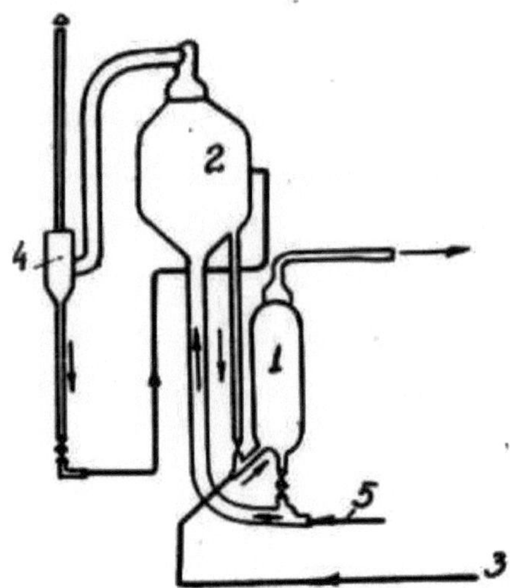

Fig . 2. Schematic and technical diagram of the installations.

Model 2. 1-Reactor, 2-Regenerator, 3-Raw materials 4-electrofilter, 5-

air

A characteristic feature of these installations was the introduction into the practice of catalytic cracking of the hydrodynamic regime of a fluidized bed of finely dispersed catalyst with downward flow of the catalyst. In other words, in these installations the gas velocity in the reactor decreased, as a result of which the solid phase concentrated in the gas flow, forming a dense turbulent layer with a pronounced upper level. Thus, here only a small part (approximately 5%) of the total amount of circulating catalyst enters the cyclones for collecting dust and the main part of the catalyst is in the apparatuses /2/.

As a result of moving from top to bottom the place where the catalyst is removed from the reactor and regenerator and simultaneously reducing the speeds in these devices, the catalyst circulation circuit was simplified and the hydraulic resistance of the system was reduced.

Both of these made it possible to reduce the total height of the installation to 50-55 meters, reduce metal consumption, and reduce capital costs and operating costs. However, the installation height was still high.

Improvement of catalytic cracking units in order to reduce their initial cost and operating costs by reducing the height led to the creation of a type III unit , the diagram of which is shown in Fig. 3.

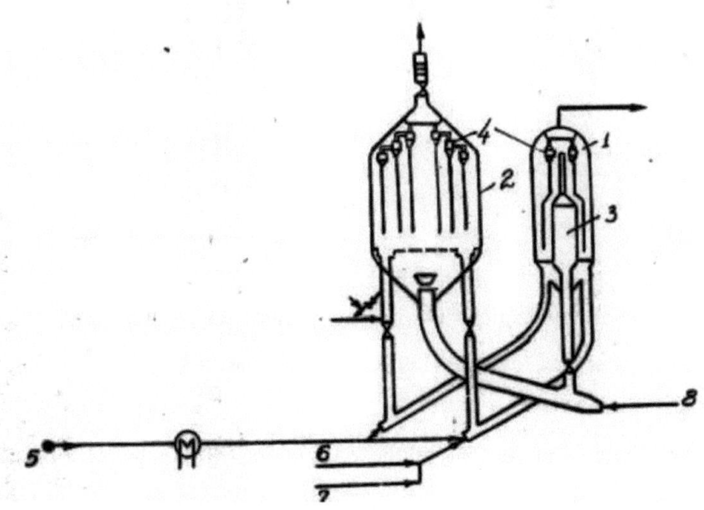

Fig.3. Principal technical diagram of the installations.

Model 3 1- Reactor , 2-Regenerator, 3-Sexion of vapors, 4-Cyclones, 5-Raw materials, 6-water steam, 7-Risauk, 8-air

The characteristic difference of this installation is the location of the reactor and regenerator on the same level, the reduced volume of the regenerator and the fact that it has a lower height of 37-40 m. All this was achieved by slightly increasing the pressure in the regenerator.

In the series of model installations, the last are installations model IV , the schematic diagram of which is presented in Fig. 4.

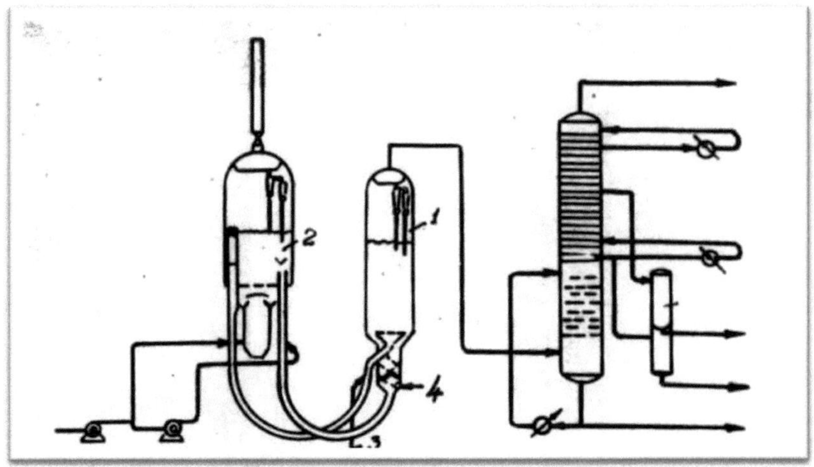

Fig. 4. Schematic and technical diagram of the installations.
Model4. 1- Reactor, 2- Regenerator, 3- Raw materials 4- air

In terms of the arrangement of the equipment, Model IV resembles Model III .

The transport of the catalyst in a dense layer through U -shaped catalyst pipes made it possible to give the "Model IV " type installations all the advantages of the "Model III " type installations without increasing the pressure of all the air supplied to the regenerator.

IV units, the intensity of catalyst circulation between the reactor and the regenerator is controlled mainly by changing the equality of flow densities

48

by supplying more or less air to the upper section of the spent catalyst pipeline.

III units , and fluidized layers in the apparatus are formed at the maximum permissible speeds of gas-steam flows.

In addition to the model series cracking units discussed above, other types of units have also been used industrially. First of all, we should include installations such as Ortoflow and Unionoil Products (UOP). /2/

Both of these types of installations are distinguished by the use of a combined apparatus, in which the reactor and regenerator are enclosed in a common housing, and installations of the Ortoflow type have two models - "A" and "B", equal in the relative position of the reactor and regenerator.

In fig. Figure 5 shows a diagram of the installation of Ortoflow model "A", and figure 7 also shows model "B"/2-24/.

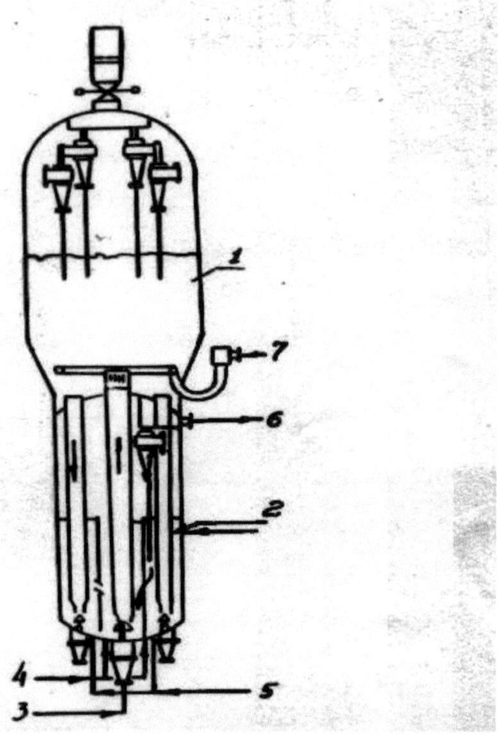

Fig.5. Fundamental technical diagram of catalytic cracking
installations orthoflou "A" type
1-Reactor, 2-Regenerator, 3-air, 4-Steam, 5-Feedstock, 6-catalyst, 7-air

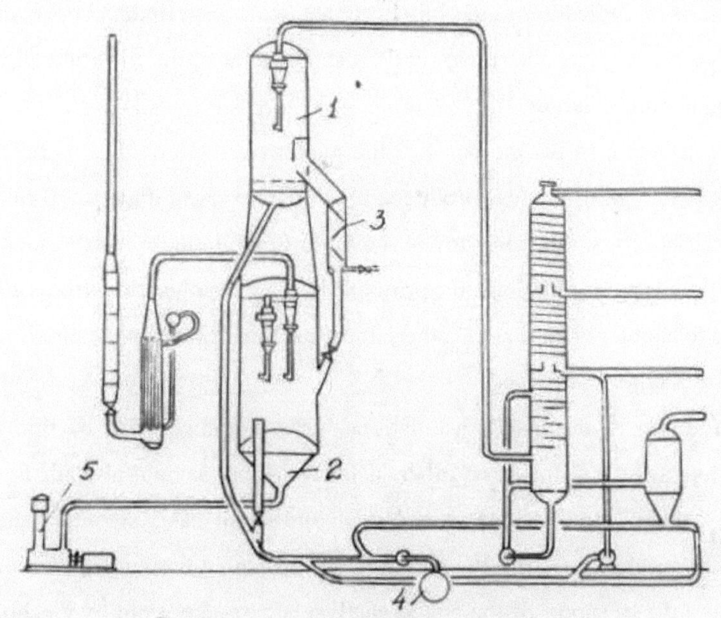

Fig.6. Basic technical diagram of installations

type YuOP.1-Reactor, 2-Regenerator, 3-Stripping Section, 4-Raw Materials, 5-Air

Considering both schemes, one should note the original structural layout and compact design. The main advantage of this type of installation is the use of a vertical (without bends or turns) catalyst pipeline, which minimizes abrasive wear of the equipment.

In addition to the original design solution of the overall system layout, Ortoflow plants differ from other plants by the presence of a direct flow of catalyst without the use of a driving agent from the reactor to the regenerator for model "A" and from the regenerator to the reactor for model "B". /24/

There is an indication in the literature that, from a mechanical point of view, it is easier to place the reactor under the regenerator, since the diameter of the regenerator is larger.

This principle forms the basis of the modern version of the fluidized bed catalytic cracking process produced by UOP. A typical diagram of one of the installations of this company is shown in Fig.6. In these installations, like Ortoflow-type installations, the principle of direct catalyst flow from one apparatus to another is retained. At the same time, the lifting catalyst pipeline in these installations is curved.

In addition to the listed installations with a fluidized bed of finely dispersed catalyst, a number of other installations are known abroad. For example, there is an installation in which the spent catalyst enters the regenerator not directly from the reactor , but through a pressure separator bunker, and the transport of the coked catalyst is carried out not by air, but by water vapor.

However, due to the fact that their total number is small and they represent only a modification of the installations described above, we will not dwell on them. At the same time, we note that we deliberately did not touch upon the industrial installations, although few, described in the literature, in which there is a tendency to intensify the catalytic cracking process. These are installations for 2-stage cracking and with a sectioned fluidized bed. These settings will be discussed below./2-24/

At the end of the twentieth century, ongoing research in the field of catalysis established a relationship between the surface properties of celite-containing catalysts, coking and the influence of this factor on the activity and selectivity of the catalytic cracking process. A distinctive feature of the conducted research was the study of these patterns for the first time in seconds of contact of the raw material with the catalyst.

The results obtained made it possible to model the catalyst according to the height of the reactor and to regulate the surface properties of the catalyst by its coking. This became the scientific basis for creating a technology for processing unstable gasoline and was confirmed during the processing of pyrocondensate at the G-43-107 catalytic cracking complex /7,8,23/.

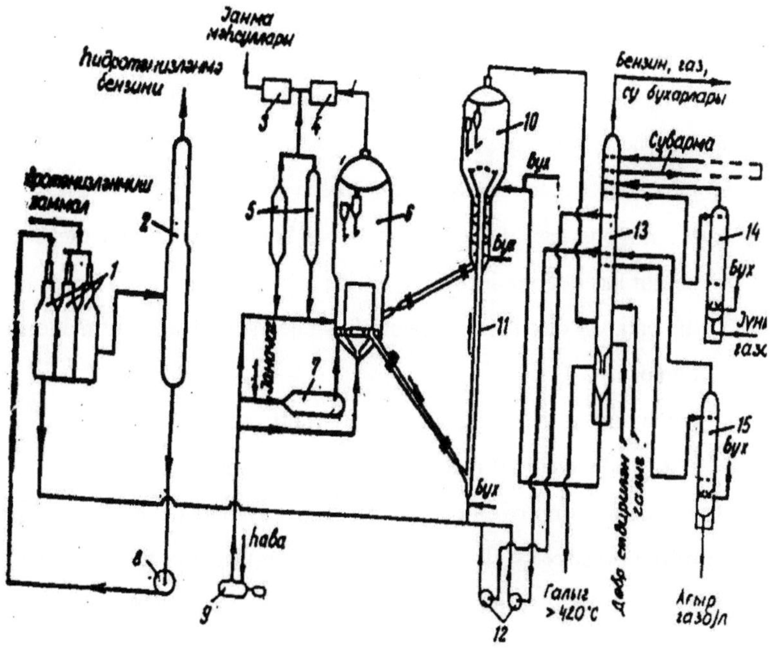

Figure 7. G-43-107 catalytic cracking unit

The table presents data characterizing the distribution of some of the installations described above in the USA, their technological indicators, characteristics of raw materials and resulting products.

Summarizing the above, it becomes obvious that since the launch of the first installations, the design of devices and installations as a whole has been significantly improved. As a result of these improvements, installations have become significantly simpler and more compact.

Ultimately, the need for metal for the construction of installations and their cost are currently approximately three times less than for the first catalytic cracking installations.

The main improvement of the installations was aimed mainly at reducing the cost of the process by reducing capital investments and consisted of the following:

1. the catalyst recovery system has been simplified;
2. the dimensions of the devices have been reduced and the overall height has been reduced;
3. the catalyst circulation system has been simplified and made more compact;
4. the layout of the devices has been improved;
5. The productivity of installations has been increased.

However, this improvement did not have any significant impact on the essence of the process itself and the yield of the target products (as can be seen from the table), therefore, along with the indicated improvement of the installations, a lot of work was carried out on the further development of the catalytic cracking process. The most important moments in the development of the process were work on deepening the process, involving high-boiling gas oil fractions for cracking , developing new types of catalysts, etc. At that time, as a result of summarizing a large number of scientific research works and accumulated experience in the operation of installations with a fluidized bed of a catalyst, serious shortcomings were identified , characteristic of this method of contact./ 24/

It was necessary, based on an in-depth study of the catalytic cracking process occurring in hollow contactors using a fluidized bed of finely dispersed catalyst, to outline ways for further intensification of the process.

5. *THE ROLE OF AZERBAIJANI SCIENTISTS IN THE DEVELOPMENT OF THE CATALYTIC CRACKING PROCESS*

The first industrial catalytic cracking unit was a low-productivity unit of the Goodry system, commissioned in the USA (1936). The industrial development of catalytic cracking in the USSR began with the construction of Goudry units. Later, more advanced catalytic cracking systems based on mobile granular and finely dispersed catalysts were developed. /4/

The process of catalytic cracking with a mobile granular catalyst in the USSR was developed by teams from VNIINP, LenNII, GrozNII and GIPROneftezavod. In the same period (1944-1952), teams of Azerbaijani scientists from AzNIINP, GiproAzneft, and factories of the Azneftezavody association developed a more effective version of catalytic cracking in a "fluidized" layer of a finely dispersed catalyst and synthesized a microspherical catalyst. All stages of the process development are reflected in the relevant reports of AzNIINP and in the project materials of GiproAzneft, which are almost not covered in the press and are little known to a wide range of readers.

At AzNIINP, research on catalytic cracking in a fluidized bed with a finely dispersed catalyst began back in 1944.

The first model in the USSR of catalytic cracking with a "fluidized" layer of a finely dispersed catalyst was implemented in 1945, and on this model the main parameters of the "fluidized " layer of catalyst particles in a flow of vapors and gases were studied, which revealed the great efficiency of this contacting method. /1/

The initial data obtained during the development of a semi-factory catalytic cracking unit in a fluidized bed of finely dispersed catalyst, built by

Baku specialists, served as the basis for the design and construction of the first industrial unit of type 1-B at the Novobakinsky oil refinery (now PA "Azerneftyanajag"), which had a huge impact on the development of the USSR oil refining industry as a whole .

Using the example of an industrial installation for catalytic cracking of kerosene-gas oil fraction for the production of aviation gasoline, put into operation in 1952, the high efficiency of systems with a "fluidized" catalyst bed was confirmed in comparison with industrially operated installations with a moving ball catalyst type 42-103, developed by GrozNII, VNIINP and other organizations. This served as the basis for the construction throughout the USSR, starting from the 50s, of a large number of catalytic cracking units in a fluidized bed of high productivity type 1-A and 1A/1 M using INHP technology and made it possible to solve the problem of - producing high quality gasoline in the country. Similar installations have also been built in Poland and Romania. /1-22/

A major contribution to the development of the theoretical foundations of catalytic processing of petroleum feedstock using fluidization technology was the research carried out under the leadership of V.S. Gutyri, R.G. Ismailov, V.S. Aliev, A.D. Lemberansky, D. A. Guseinov, V.P. Kramskoy, V.V. Manshilin, N.M. Indyukova, A.M. Kuliev, M.I. Rustamova, Sh.K. Kyazimov, I.I. Sidorchuk, A.G. Bagirovya F I Zeynalov and many others.

Along with the development and industrial implementation of the catalytic cracking process with a fluidized bed of aluminosilicate catalyst, M.I. Rustamov conducted extensive research to study the structure of the "boiling" layer, the patterns of mass and heat transfer properties of the fluidized layer, and its hydrodynamics . /1,7,8,22/

As a result of these studies, some provisions of the two-phase model of the "fluidized" bed were refined and an equation was derived that makes it possible to determine the amount of a discrete phase depending on the properties and size of solid particles forming the "fluidized" bed and the fluidization number.

It has been established that intense mixing of the solid phase occurs in the "fluidized" bed. As a result, the temperature in the volume of the layer is equalized, which ensures that the process proceeds, regardless of the magnitude of the thermal effect of the reactions, under isothermal conditions. Intensive mixing of solid particles, on the other hand, leads to averaging of the activity of the catalyst in the reaction volume and helps to reduce the concentration of the reacting components as a result of their dilution with reaction products involved in intra-reactor circulation with the catalyst . /1,2,13/

Both the formation of bubbles and the mixing of the gas-catalyst mass in the volume of the layer lead to a significant decrease in the rate of the reactions occurring and the productivity per unit of reaction volume.

The presence of these shortcomings does not allow full use of the potential possibilities of contacting the gas and solid phases.

In this regard, extensive research has been carried out, aimed, firstly, at eliminating these disadvantages of fluidized bed systems and, secondly, at developing new contacting principles.

Supporters of the first direction followed the path of improving systems with a "fluidized" bed. This was achieved by dividing the volume of the reaction apparatus into sections or stages, which made it possible to limit the intense mixing of solid particles and gas within small volumes, i.e. within one section or step. In this case, although in each stage or section the disadvantages of the "fluidized" layer remain, but, in the presence of several successively located sections, gradients of catalyst concentrations and activity are created between them /22/.

With an infinitely large number of sections , the disadvantages inherent in the "fluidized" bed are completely eliminated, and the apparatus mode approaches the ideal displacement mode.

, teams from GrozNII (with direct-flow phase movement) and VNIINP (with counter-flow -step phase movement) worked towards the creation of reactor devices with a sectioned "fluidized" bed.

Unlike these institutes, the Institute of Chemical Industry of the Academy of Sciences of Azerbaijan was the first to develop a fundamentally new method of contacting solid and gas phases, which makes it possible to subsequently create progressive catalytic cracking systems using the developed principle of contacting in an ascending through and semi-through catalyst flow. These studies have developed since 1957 under the leadership of B. S. Aliyeva and M. I. Rustamova.

The method of contacting gas and solid phases in systems with an ascending through and semi-through flow, as well as the method of implementing various contact-catalytic processes in these systems, is protected by several copyright certificates. The creation of new systems for contacting phases in an ascending through and semi-through flow, allowing processes to be carried out under ideal conditions displacement was preceded by theoretical studies conducted by M.I. Rustamov to determine the efficiency of sectioning systems with a "fluidized" bed at various values of the fluidization number for irreversible monomolecular reactions *.)*

As a result of these studies, it was found that a practically noticeable increase in the efficiency of the process is achieved by increasing the number of sections to 6-8. At the same time, the efficiency of the device increases by 15-20%. A further increase in sections increases the efficiency of the apparatus so insignificantly that a further increase in their number becomes ineffective. /7,8,22/

Subsequently, when comparing data from pilot catalytic cracking units with reactor devices with a sectioned "fluidized" bed under co-current conditions (GrozNII), step-countercurrent systems (VNIINP) and systems with an ascending flow, the theoretical conclusions were basically confirmed, and the advantage of the systems was revealed with an upward flow.

The results obtained became the basis for creating a method for theoretically assessing the effectiveness of various contact methods.

An equation was derived that makes it possible to determine the efficiency of a particular contacting system for carrying out chemical reactions, based on the number of collisions of fine catalyst particles with substrate molecules per unit volume of the reactor per unit of time, and a theoretical assessment was given of the efficiency of reactor devices operating on the principle of phase contact in "fluidized" layer, in ascending through and semi-through flows.

The thermal properties of systems with a circulating finely dispersed catalyst between a coupled reactor and a regenerator are considered, in one of which, as a result of an endothermic chemical process, coke deposits are formed along with the target products, in the other, as a result of the chemical reaction of burning coke deposits, heat is released, which with the circulating catalyst is introduced into the reactor to compensate for the latter's need for thermal energy. Calculation equations have been compiled that make it possible to determine the balanced thermal regime in the devices under conditions of their optimal operation by changing the amount of circulating catalyst. /2-22/

Contacting the solid and gas phases under conditions of ascending through and semi-through flows ensured that the process was carried out in a regime close to ideal displacement. The created reactor designs made it possible to increase their specific productivity by 8-10 times compared to a fluidized bed. /22-23/

The use of systems with an upward flow of catalyst also made it possible to increase the productivity per unit mass of the catalyst several times, significantly reduce coke formation, and increase the selectivity of the catalytic cracking process at the same depth of raw material conversion. The possibility of reducing coke formation made it possible, using these systems, to develop processes for catalytic cracking of heavier raw materials, the cracking of which could not be carried out in systems with a "fluidized" bed.

These advantages predetermined the high economic effect obtained from the implementation of this system in industry.

Research into catalytic cracking in an upflow system was completed in 1960. In the same year, several US patents appeared on a method for carrying out the catalytic cracking process using up-through-flow reactors.

For the period 1961-1964. Using systems with an upward flow of catalyst , various modifications of two-stage catalytic cracking of vacuum stripping were developed with the first stage carried out in an upward flow, the second in a semi-through flow in selective and non-selective options.

As is known, one of the main factors determining the yield of reaction products, especially gasoline and coke, in two-stage catalytic cracking is the degree of conversion of raw materials in stages I and II of the process. /2,22,23/

With the same degree of conversion of raw materials in two stages, the total yield of individual reaction products can be different depending on the change in the degree of conversion of raw materials in each stage.

Depending on the degree of conversion of the raw material in stage I , the significance and contribution of stage II to the total yield of the target product is determined. To this end, under the leadership of Rustamov M.I., Pryanikova E.I., Abdullaev M.A., systematic, targeted research was carried out to establish the influence of the depth of the 1st and 2nd stages and, as a consequence, the quality of recycling on the efficiency of the process as a whole.

These studies were carried out according to a two-stage selective and non-selective scheme for processing fresh raw materials and recycling with separate and joint removal of reaction products at each stage of the process.

As a result of the studies, it was established that of the three schemes studied for the implementation of the process of catalytic cracking of vacuum stripping, the highest yield of gasoline with less coke formation is provided by the scheme of two-stage catalytic cracking with separate removal of products of stages I and II. The lowest gasoline yield with relatively high coke formation is ensured by a single-stage joint catalytic cracking scheme for vacuum stripping and recycling. The scheme of two-stage catalytic cracking with the joint removal of products of stages I and II occupies an intermediate position.

It has been established that with a decrease in the amount of recycling returned for repeated cracking, the indicators of the three comparative schemes come closer, and when the amount of recycling is less than 10% of the mass, the scheme of single-stage joint cracking of raw materials and recycling is likely to be the most effective, since it has simpler hardware design.

As a result of studies of the influence of the depth of conversion of raw materials at stages I and II of the process, it was shown that, depending on the depth of transformation in stage I, the resulting refluxes differ in quality and have different cracking abilities. An increase in the depth of conversion of raw materials in the first stage leads to an increase in the content of aromatic hydrocarbons, resins and asphaltenes in the resulting reflux, which contributes to a decrease in the yield of gasoline and an increase in the yield of coke in the second stage of the process.

It has been established that the optimal depths of conversion of raw materials in stages I and II of the process, which ensure the maximum total yield of gasoline in two-stage catalytic cracking with a lower yield of coke, differ from the depths of conversion of raw materials that provide the maximum yield of gasoline of stages I and II separately . The optimal depth of conversion of raw materials in the first stage of the process in terms of the total indicators of two-stage cracking is in the middle zone, and in our studies it was determined in the region of 50-65% by weight.

A comparative assessment of the quantitative and qualitative indicators of the process on zeolite containing and amorphous aluminosilicate catalysts is given.

It has been determined that the use of a zeolite containing catalyst provides a better material balance and selectivity of the process, a slightly better quality of the resulting gasoline, a decrease in the content of unsaturated hydrocarbons, an increase in the content of isomeric hydrocarbons in the gas composition, as well as an increase in the content of aromatic hydrocarbons in liquid products, especially gas oils, which are the raw materials. to obtain soot concentrate.

In June 1963, the results of research by the Institute of Chemical Industry of the Academy of Sciences of Azerbaijan on the development of catalytic cracking systems with an upward flow of catalyst were first reported at the All-Union Meeting in Ufa.

In 1966, the first monograph devoted to this issue was published (Aliev B.S. , Rustamov M.I., Pryanikov E.I. - " Current state and ways of intensifying the catalytic cracking process " Baku, Azerneshr 1966) /2/, where the scientific foundations of the catalytic cracking process in systems with rising through and semi-through flow, and the indicators of the catalytic cracking process in systems with rising flow of various modifications were outlined.

This contributed to the widespread use of these systems for the development of the catalytic cracking process in such scientific organizations as GrozNII and VNIINP, which had previously proposed and developed systems with sectioned apparatus.

Based on the research conducted by INHP scientists, recommendations were issued and the reconstruction of industrial catalytic cracking units with fine catalysts was carried out at a number of refineries using upflow and semi-through flow reactors, which made it possible to sharply increase the production of high-quality motor gasoline. /7-8/

For the development and introduction into industry of the process of two-stage catalytic cracking of petroleum raw materials in 1982, V.S. Aliyev, M.I. Rustamov, E.I. Pryanikov were awarded the State Prize of Azerbaijan./22/

In parallel with the development of catalytic cracking technology in an ascending gas catalyst flow, the Institute of Chemical Petroleum Production of the Academy of Sciences of Azerbaijan carried out extensive work to create the scientific basis for processing petroleum feedstock in such systems. The structure of the flow has been studied and the hydrodynamic characteristics of the devices have been assessed within a wide range of changes in the linear velocity of the medium and the fluid concentration of solid particles. Transition regions from the "fluidized" layer to the semi-through layer and then to the through flow have been established.

As a result of the fact that in systems with an ascending flow, a high relative speed of phase movement is ensured with intense transverse and local- longitudinal oscillatory motion of particles, and also, taking into account the high dispersion of particles (40-100 c) and the possibility of flow of reacting particles around each particle molecules, we can assume that the process in these systems occurs in the intraphase kinetic region. This is especially convincing for slow processes, such as catalytic cracking.

Based on the studies carried out, a mechanism for the process of catalytic cracking of petroleum feedstock in reactors with an upward flow has been proposed and a kinetic model of the process has been compiled.

The presence of kinetic and hydrodynamic models and heat balance equations allowed M.I. Rustamov, R.I. Zeynalov and G.T. Farkhadova to develop a mathematical model of the process and solve the optimization problem under unsteady conditions of the process.

Later, using rising flow reactors , together with GrozNII and VNIINP, a new catalytic cracking complex G-43-107 was developed, characterized by higher productivity and low metal consumption. This is a combined unit, which includes sections for hydrotreating vacuum gas oil, catalytic cracking of hydrotreated vacuum gas oil in a vertical lift reactor, and gas fractionation

.

It should be noted that this system is the most effective because it allows you to identify all the advantages of zeolite-containing catalysts, provides high yields of gasoline, propane-propylene and butane-butylene fractions, which makes this installation competitive at the international level (a license has been sold for the process).

Currently, G-43-107 units are successfully operated at oil refineries in Moscow, Grozny, Mozhikiai, Burgas, Ufa, Pavlodar, Angarsk, Lisichansk, and Baku. Over the three years of operation of the G-43-107 complex in Baku, contributions to the state budget amounted to more than 750 billion manats. According to European experts, at present the installations are at the level of foreign analogues and are the only ones at refineries of the former USSR that do not require serious modernization. /2-22/

Simultaneously with the introduction of new, progressive catalytic cracking systems in the period 1970-1975. Intensive studies are being carried out on the patterns of catalytic cracking in a wide temperature range (400-800°C). A simultaneous study of the catalytic transformations of hydrocarbons in such a wide temperature range using the same catalyst has not been carried out until now. Carrying out such studies made it possible to consider from a scientific perspective the change in the mechanism of the process in a wide temperature range, and from a practical perspective to identify its selectivity in the yield of individual hydrocarbons in the composition of the reaction products.

Analysis of the component composition of the target products showed that it is consistent with the radical carbonium ion mechanism; with increasing temperature, the proportion of reactions proceeding by the radical mechanism increases.

In the region above 600°C, a sharp qualitative change in the composition of the reaction products is observed. The observed distribution of hydrocarbons in the final product indicates a change in the mechanism of the process and its transition to the area of catalytic pyrolysis. The transformation of hydrocarbons in this temperature range apparently obeys a radical mechanism. The catalyst does not change the radical chain mechanism of the process. Its role probably boils down to initiating the formation of additional radicals by a heterogeneous surface and shifting the optimal yield of olefins to lower temperatures than during thermal cracking.

As a result of studies of the role of operating conditions and the molecular weight of the initial petroleum feedstock in the formation of the structure and molecular weight of the final products, it became possible to identify the area of catalytic cracking in which the yield of low molecular weight isoolefin and isoparaffin hydrocarbons is maximized. These studies open up new aspects of the development of the catalytic cracking process. At the same time, in contrast to the traditional purpose of the catalytic cracking process, the intended purpose of which is the production of motor gasoline, its new modification - the process of destructive isomerization is considered as the main process of the petrochemical complex, the main purpose of which is the production of low molecular weight isoparaffin and isoolefin hydrocarbons.

The process is called destructive isomerization, due to the fact that it is not the original hydrocarbon that undergoes isomerization , but the actual product of its cleavage (destruction). As a result of a study of the patterns of destructive isomerization of technical raw materials of primary and secondary origin, it was shown that various oil fractions from gasoline to vacuum distillates can be involved in the process. Coking distillates can be involved in the process as one of the types of raw materials. /22/

A study of the influence of hydrodynamic conditions of the process showed that when the process is carried out in reactors with an upward flow under conditions of approximately the same conversion depth, the yield and selectivity of the process for the target product is higher than when it is implemented in reactors with a fluidized bed of a catalyst.

The destructive isomerization process has been developed in two versions: petrochemical and fuel. The petrochemical option is aimed at producing low molecular weight hydrocarbons and hydrocarbons, mainly i -C $_4$ and İ-C $_{5\ olefins}$. The yield of the propane-propylene fraction, consisting mainly of propylene, is 11-12% . At the same time, the process produces ~25 wt.% of the raw material high-octane component of automobile gasoline , having an octane number of 85-89p. according to the motor method in its pure form.

The fuel option is aimed at obtaining either a high-octane isocomponent of motor gasoline or at producing high-octane gasoline. When it is sold, 30-

35% of the high-octane isocomponent is obtained, having an octane number of 86-88p. according to the motor method in pure form or ~50% of motor gasoline with an octane number of 83 pp. according to the motor method in pure form .

The fundamental basis of the proposed process scheme is the processing of fresh raw materials in reactors with an ascending flow of zeolite containing catalyst ending in a semi-through flow.

Based on the petrochemical version of the process, a comprehensive scheme for processing petroleum fractions with the main process of destructive isomerization has been developed in order to organize large-scale production of isoprene rubber.

The USSR author's certificate for this process was received in 1975. / 13/ For comparison, I would like to note that the leading US companies in the field of catalytic cracking received a patent for a similar process only in 1990-1992.

As a result of studying the role of the structural-surface properties of the zeolite catalyst, it was revealed that the destructive isomerization reaction occurs most selectively on "U" zeolites with the participation of B- and L-centers, the predominant contribution to the process is made by "strong" and "medium" B- forces centers. Apparently, the splitting reaction initially occurs at L centers. In this case, at "weak" L -centers containing $Al^{+e\ ions}$ in cluster states, the cleavage of middle C-C bonds predominantly occurs. Breaking of terminal C-C bonds is possible with the participation of "strong" L - centers./ 14,15.16/

The resulting olefins undergo further transformations at B-centers. The participation of B centers in initiating the formation of carbonium ions cannot be excluded, since the destruction of paraffins occurs in the presence of olefins formed during thermal decomposition with the participation of one-electron donor and acceptor centers. Isomerization of secondary carbocations is energetically favorable and occurs predominantly with the participation of B-centers of "average" strength. A satisfactory linear correlation has been established between the concentrations of isomerization products and active centers.

In order to identify the influence of the nature of the matrix, the conditions of its formation and the influence on the activity and selectivity of hybrid catalysts, their pore-structural characteristics and acidity spectrum, the activity of "U" type zeolite introduced into the composition of various metal silicates was studied. The relationship between the evolution of acid sites and changes in the nature and concentration of the initial components, as well as their influence on the formation of the specific surface area and activity of the catalyst, has been revealed. /15-16/

These studies were carried out under the leadership of M.I. Rustamov, G.T. Farkhadova, and during this period, in the processes of catalytic cracking and destructive isomerization, INHP was recognized as the leading organization not only in the USSR, but also in the countries of Eastern Europe within the framework of cooperation between the Academies of Sciences socialist countries, and academician Rustamov M.I. was elected international coordinator./ 22-23/

Promising environmental requirements for motor gasoline have stimulated the development of a petrochemical version of catalytic cracking, and now all over the world the question of repurposing part of catalytic cracking for the production of low molecular weight hydrocarbons is being raised, and in Finland an oil refining complex has already been launched with the operation of a catalytic cracking unit in the destructive isomerization mode.

Research into the intensification of the catalytic cracking process in order to improve the quality of the produced gasoline led to the creation of a technology for producing unleaded motor gasoline AI-93 in the catalytic cracking process itself./ 20-21/

The essence of the developed technology is a two-stage catalytic cracking of both sulfur and non-sulfur vacuum distillation with independent withdrawal and separation of the products of each stage and supply of the non-sulfur fraction to the second stage. - 195 or n.k. - 270°C from the first stage of the process.

The creation of the process was preceded by a study of its scientific foundations - detailed studies of the patterns of transformation of various classes of hydrocarbons in the composition of catalytic cracking gasoline (Stage I) and the mutual influence of their quantitative ratio on the formation of the total octane number. The work was carried out under the supervision of M.I. Rustamov and A.D. Guseinova. /20-21/

Due to the fact that, depending on the structure and content of organosulfur compounds, petroleum products were subjected to desulfurization to varying degrees, studies were carried out on the mechanism for removing various classes of sulfur compounds from gasoline. As a result, the main factors of gasoline desulfurization during their catalytic upgrading were established . In particular, it has been established that the decomposition of sulfur compounds and their redistribution between the products of catalytic upgrading proceed more efficiently if gasoline contains aliphatic sulfur compounds. Increasing the temperature of the process, in the case of mercaptan sulfur in gasoline, contributed to better desulfurization of gasoline, and in the case of containing other classes of sulfur compounds (sulfides, disulfides, thiophene, etc.), on the contrary, it made desulfurization more difficult.

The data obtained are explained by a two-stage mechanism for the removal of sulfur compounds from gasoline under conditions of their contact with the surface of an aluminosilicate catalyst.

As a result of studies of catalytic cracking of sulfurous vacuum stripping, an optimal mode was selected that ensures the production of high-octane gasoline AI-93 with the required sulfur content, according to GOST, without the use of hydrotreating. /21,22/

Due to the promising nature of the coking process, much attention has been paid to the possibility of efficient processing of coked gasoline in the process of catalytic upgrading. As a result of the research, the possibility of increasing the octane number of coking gasoline by 10 points was identified in the process of its upgrading on a zeolite-containing catalyst and producing commercial gasoline A-76 on its basis. The possibility of involving coking gasoline in the preparation of AI-93 gasoline during the catalytic refinement of a mixture of coking gasoline and catalytic cracking, taken in a ratio of 1:6, was also established.

The possibility of obtaining AI-93 gasoline using the developed technology from vacuum distillates of oils from various fields (Mangyshlak, Turkmen sulfurous, etc.), as well as through direct catalytic cracking of fuel oil and crude oil, has been confirmed.

With the deterioration of the environmental situation, the requirements for the quality of gasoline continue to become stricter.

New requirements are associated with the need to change the chemical composition of gasoline and produce gasoline that contains a limited content of aromatic and unsaturated hydrocarbons and provides for an increase in the proportion of oxygen-containing compounds.

Research in this direction, conducted under the leadership of M.I. Rustamov, G.T. Farkhadova and A.D. Guseinova , led to the creation of a combined process of catalytic cracking and low-temperature upgrading of secondary gasolines by direct contact of their narrow fractions with monohydric alcohols on zeolite-containing catalysts. It was shown for the first time that when carrying out the process on zeolites, in parallel with the formation of a homologous series of tertiary alkyl ethers, reactions of structural and skeletal isomerization of n-olefins, alkylation, i.e., occur. In the composition of oxygenated gasoline , the content of hydrocarbons with higher octane characteristics increases . /22/

The reaction proceeds according to the carbonium ion mechanism with the participation of B- and L -centers, and the contribution of L -centers to the total catalytic activity is significantly higher and amounts to 70-80 %.

The resulting esters are homogenizers and allow significant amounts of free methanol to be included in the gasoline composition.

It should be noted that, unlike all other processes of a similar purpose, in which an improvement in the quality of gasoline is accompanied by a decrease in their resources, the process of low temperature upgrading is, perhaps, the only one in which an improvement in the quality of gasoline is accompanied by an increase in their resources due to alternative raw materials.

One of the main purposes of the catalytic cracking process is to deepen oil refining. In this regard, studies have been carried out on the catalytic cracking of fuel oil and crude oil in reactors with an upward and semi-through flow of catalyst.

Work in this direction was carried out under the leadership of M.I. Rustamov, E.M. Seid-Rzaeva, G.T. Farkhadova./22/

A less capital-intensive process for catalytic cracking of oil in a reactor with an upward flow of catalyst has been studied and developed, allowing for the production of more than 70% of light motor fuels with a fewer number of technological stages.

The scientific basis for various options for processing heavy petroleum feedstocks on a finely dispersed catalyst has been developed, which ensures a high depth of their processing and the production of a significantly larger amount of light petroleum products.

Options were explored for both direct involvement of heavy oil residues in catalytic cracking feedstocks and pre-refining them on newly developed inert contact catalysts, bringing their quality to the requirements for traditional catalytic cracking feedstocks.

For the first time, new contact-containing catalysts for the processing of heavy petroleum feedstocks were synthesized and studied, the relationship between the structural characteristics of the synthesized catalysts and their activity and selectivity in the process of catalytic upgrading and cracking was established, and the influence of these factors on the chemistry and mechanism of the process was studied. It has been established that the process of converting oil and fuel oil on synthesized catalysts (Stage I) ensures their upgrading by reducing the content of nitrogen, sulfur, asphalt

-resinous substances and heavy metals (Ni, V) to the standards required for vacuum stripping . Based on this, a process for 2-stage processing of fuel oil in variants for producing gasoline and diesel fuel and a comprehensive scheme for residue-free processing of crude oil have been developed using synthesized catalysts.

For the first time, a set of studies was carried out to establish the patterns of cracking of vacuum gas oils of different depths of selection in the presence of activators and without them.

The influence of activators of various natures under optimal conditions on the intensity of individual reactions has been revealed.

Using a complex of physicochemical methods of analysis, including spectral (IR, PMR, ESR, etc.), the influence of the cyclicity of aromatic hydrocarbons contained in the composition of activating additives on the efficiency of their action in the catalytic cracking process was revealed, expressed in increasing the yield of gasoline and decreasing the yield of coke. The bifunctional properties of the synthesized antimony containing polyarylene in the process of catalytic cracking of vacuum gas oil with fuel oil were revealed. It has been established that its optimal amount, along with passivating and coke-reducing properties, helps to increase the octane numbers of gasoline and increase the afterburning of CO into CO_2 during the regeneration process. Based on the establishment of general patterns of action of activators in a complex dispersed catalytic system and the main indicators of catalytic cracking of heavy petroleum feedstock, an integrated approach to the catalyst and process feedstock with positions for regulating phase transitions in dispersed systems, which makes it possible to predict and intensify the process of processing heavy petroleum feedstock.

Numerous studies have been carried out to study the influence of activating additives and passivators on the patterns of cracking of heavy oil residues .

The effectiveness of the latter has been shown, but the mechanism of their action remains controversial.

Research in the field of hydrocracking of various petroleum feedstocks played a major role in the development of complex schemes for processing heavy petroleum feedstocks. These studies were carried out under the leadership of M.I. Rustamov and Kh.I. Abadzade.

As a result of these studies, an effective catalyst was synthesized by introducing Ni into a zeolite containing aluminosilicate, which operates stably at relatively low pressures and ensures the effective conversion of raw materials of various fractional chemical compositions.

Work in the field of hydrocracking developed mainly in the following areas:

- hydrocracking of heavy low-octane straight-run gasoline fractions and coking gasoline fractions in order to obtain a high-octane isocomponent of AI-93 gasoline without thermal power plants and raw materials for petrochemicals - isobutane and isopentane;

- hydroisomerization of catalytic cracking gasoline in order to improve stability and increase its octane characteristics;

- hydrocracking of catalytic cracking gas oils at a pressure of 7 and 1 MPa to produce gasoline, jet and diesel fuels;

- light hydrocracking under a pressure of 5 MPa of vacuum distillates of sulfur and low-sulfur oils in order to obtain summer diesel fuel and low-sulfur raw materials for catalytic cracking;

- light hydrocracking of a mixture of vacuum distillate with light coking gas oil, heavy coking gas oil , light catalytic cracking gas oil and heavy - catalytic cracking gas oil in order to obtain summer diesel fuel and low sulfur feedstock for catalytic cracking;

- light hydrocracking at a pressure of 7 MPa of heavy residual fractions of oil-fuel oil, tar (after preliminary deasphalting with propane) in a mixture with vacuum stripping in order to obtain summer diesel fuel and low-sulfur raw materials for catalytic cracking and high-index oils.

The process for producing isoparaffin hydrocarbons using the developed catalyst has received patents from the leading capitalist countries of the USA, England, France, Canada, Germany, and Japan.

The influence of the chemical composition on the results of hydrocracking of heavy raw materials was studied.

Using mass spectrometry and UV spectroscopy, it was established that under light hydrocracking conditions the reactivity of aromatic hydrocarbons is very low. Their transformation occurs mainly due to the detachment of side chains. Of these, only polycyclic compounds undergo shallow hydrocleavage. In hydrogenates, compared to the raw material, the content of alkylbenzenes often even increases due to the fact that their formation is the final stage of the partial decomposition of the saturated parts of more complex semi-hydrogenated aromatic hydrocarbons and heteroorganic compounds with a higher molecular weight.

It has been revealed that in the process of light hydrocracking, mainly destruction reactions of paraffin and polycyclic naphthenic hydrocarbons occur, as a result of which during light hydrocracking of distillate raw materials. A low content of aromatic hydrocarbons ensures the production of diesel fuel. The yield of diesel fuel in the process of light hydrocracking increases as the content of aromatic hydrocarbons and resins in the feedstock decreases, which indicates the inhibitory effect of these compounds.

In order to increase the stability of the synthesized zeolite-containing Ni-Al-Si-ro catalyst, the development of a process for light hydrocracking of heavy types of raw materials was carried out over a two-layer catalyst using an industrial hydrotreating catalyst (Al-Co-Mo) and zeolite- containing Ni - Al - as the top layer Si - ro hydrocracking catalyst as the bottom layer. Testing of a two-layer catalyst loading system in a pilot plant in the process of light hydrocracking of heavy types of raw materials for 6000 hours showed its high, stable activity.

It has been established that with light hydrocracking of vacuum stripping and its mixtures with secondary gas oil fractions on a zeolite-containing Ni-Al-Si catalyst, the yield of summer diesel fuel in accordance with GOST 4749-49 is 38-41% by weight with a residual yield of fr.>360 °C - high-quality raw materials for catalytic cracking 56-58% wt.

Based on the conducted research, schemes for deep processing of fuel oil are recommended, including light hydrocracking of vacuum stripping and its mixture with **gas oil** fractions of catalytic cracking and coking, mixtures of vacuum stripping with deasphalted tar , deasphalted fuel oil on a zeolite-containing Ni - Al - Si catalyst.

Speaking about the work carried out in the field of catalytic cracking, I cannot help but dwell on studies of the relationship between the surface properties of zeolite-containing catalysts, coking and the influence of this factor on the activity and selectivity of the catalytic cracking process. The work was carried out under the guidance of Rustamov M.I., Farhadova G.T.

, Saidova A.A., Amirbekov E.N. A special feature of the studies carried out was the study of these patterns in the first seconds of contact of the raw material with the catalyst, when it exhibits maximum activity, as well as the study of the dynamics of changes in the morphology of the resulting coke. As a result of the studies, a correlation was established between the activity, selectivity, and acidity spectrum of zeolite containing catalysts during the initial period of catalyst coking. In this case, the different nature of the change in the cracking, disproportioning and isomerizing ability of the catalyst in the initial period of coke deposition is associated with the unequal degree of blocking of B- and L -centers of varying strength by coke deposits.

Using the methods of elemental analysis, EPR, IR spectroscopy and electron microscopy, it was revealed that in the process of coke deposition, "strong" acid sites responsible for decomposition and coke formation reactions are most intensively blocked, the concentration of "medium" acid sites responsible for isomerization reactions, decreases with coke deposition much more slowly, and "weak" acid sites, mainly responsible for spatial isomerization, are present on the surface of the catalyst even when it is completely filled with coke deposits.

As coke accumulates on the catalyst, the concentration of paramagnetic centers and the C:H ratio increase , i.e. dehydration of coke, which is a consequence of the disproportionation of hydrogen between coke molecules and reaction products. Thus, during the coking process, coke, being depleted of hydrogen, simultaneously supplies it to the reaction zone.

That is why for disproportionation reactions to occur, a certain content of chemisorbed compaction products on the catalyst surface is required.

The results obtained made it possible to model the catalyst along the height of a once-through reactor with an upward flow of catalyst and to regulate the surface properties of the catalyst by its coking.

As a result of studying the morphology of coke deposits, it was established that as the time of contact of the catalyst with the hydrocarbon increases, stable aromatic compounds with a higher molecular weight grow in the composition of coke deposits, represented in the initial period of coke deposition by stable polyene structures and less stable compaction products of an aromatic nature. Differences in the genesis of coke deposits were revealed, associated with the occurrence of coke formation by two mechanisms: dehydrogenation of the original molecule at L -centers and the mechanism of protonation of olefins at "strong" B-centers.

Reprints of the publications of this work were requested by all developed countries of the world. /22/

The results obtained became the scientific basis for the creation of a technology for processing unstable gasoline on a partially coked catalyst in the process of catalytic cracking. These studies were carried out by M.I. Rustamov, G.T. Farkhadova and E.R. Babaev.

It has been established that contacting unstable gasolines with a partially coked catalyst ensures their efficient processing without affecting the main operating parameters of catalytic cracking.

During the upgrading process, the amount of unsaturated hydrocarbons in gasoline is reduced due to hydrogen redistribution reactions. Structural and spatial isomerization of paraffin and olefin hydrocarbons leads to an increase in their composition of hydrocarbons with a higher octane characteristic.

The identified patterns were confirmed during the processing of pyrocondensate at the G-43-107 catalytic cracking complex.

Much attention was paid to intensifying the process of regeneration of the coked catalyst. This was achieved by regenerating coked catalysts in the presence of promoters for converting carbon monoxide into dioxide. At the same time, harmful emissions of carbon monoxide into the atmosphere are eliminated. The main component of the promoters that have received industrial sales is platinum, an effective "oxidizing agent". Due to its great shortage and high cost, in order to reduce and subsequently eliminate platinum from the composition of promoters, studies have been carried out on the oxidative activity of platinum oxide systems and purely oxide systems based on metals of variable valence (Cu, Co, Cr, etc.) during the regeneration process coked catalyst. Using the developed method, the activity of promoters in the coupled reactions of burning coke from the surface of the catalyst and burning carbon monoxide into dioxide under dynamic conditions was determined. The influence of promoters on both processes occurring in the bulk and on the surface of neighboring catalyst particles was revealed.

This, apparently, can be explained by the "jump" of oxygen ions, unstable in the gas phase, into the coke crystal lattice, where they are stabilized, especially if we take into account that at high temperatures the rate of mutual transitions greatly increases.

Based on the nature of the catalytic action, three groups of promoters have been identified:

- promoters that burn CO into CO_2 but do not affect the rate of burning coke from the surface of the coked catalyst. They are recommended for use in the process of catalytic cracking of traditional raw materials - vacuum gas oil;

- promoters that burn CO into CO $_2$ and accelerate the burning of coke from the surface of the coked catalyst. Their use is advisable both when cracking vacuum gas oil and vacuum gas oil with an increased boiling point. In the first case, the use of promoters will increase productivity and eliminate the supply of sludge to the regenerator. In the second case, it will allow the processing of heavy raw materials on existing industrial installations without their reconstruction;

- promoters that increase the burning of coke from the surface of the coked catalyst and increase the CO content in the flue gases compared to the unpromoted sample. The use of these promoters is recommended when processing heavy petroleum feedstocks. In this case, the required depth of catalyst regeneration is achieved and, due to the removal of a large amount of CO by flue gases, a normal, balanced temperature regime of the regenerator is ensured without installing any heat removal systems in it. In this case, carbon monoxide should be burned in remote catalysts.

Throughout this time, much attention has been paid to the creation of new methods and laboratory facilities, their unification and replication.

I would like to note that created by acad. M.I. Rustamov developed a laboratory pilot installation that allows one to study the laws of the catalytic cracking process under the dynamic conditions of a conjugate reactor and regenerator with small catalyst loads, an analogue of which exists only in the USA.

The Institute's work in the field of catalytic cracking is protected by numerous copyright certificates, published in the republican and foreign press, and reported at international symposia and conferences .

These results were obtained mainly by employees of the institute, who subsequently grouped together in the laboratory of catalytic cracking and pyrolysis, organized in 1967 by M.I. Rustamov.

The catalytic cracking process, being one of the oldest oil refining processes, at the same time remains one of the youngest, constantly developing and improving processes, the comprehensive study of which opens up new ways of its improvement and use.

In a brief essay, I tried to highlight the "yesterday" and "today" of catalytic cracking. What is the "tomorrow" of this process, the possibilities of which are inexhaustible?

In the future, the role of the petrochemical option will undoubtedly increase, and part of catalytic cracking will be repurposed for the production of low molecular weight hydrocarbons.

One of the problems will be the involvement of heavier types of raw materials in the process, which require the development of special multifunctional catalysts and carrying out the process under millisecond conditions.

Since "qualitative" leaps in the development of the catalytic cracking process are inextricably linked with the use of qualitatively new catalysts, much attention of researchers will be paid to the creation of new catalysts. One of the main requirements for them will be to reduce the energy intensity of the process.

Currently, research is being carried out on a new generation of catalysts obtained by depositing metal complexes on the surface of aluminum oxide by a cryochemical method. Catalysts of this type significantly reduce the temperature at which C–C bonds are destroyed and make it possible to avoid aggregation of active centers on the surface. Low-temperature destruction of hydrocarbons and technical petroleum fractions in the presence of hydrogen at normal pressure was carried out using supported metal catalysts operating for more than one year without regeneration. The use of these catalysts will make it possible to create a fundamentally new generation of catalytic cracking technology.

Much attention will be paid to the environmental friendliness of the process. From this position, the research carried out at the Institute into the laws of the process in the presence of carbon dioxide, which plays the role of both a weak oxidizing agent and a transport agent in the process, is of great interest.

One aspect of process improvement will be the efficient use of all products produced by catalytic cracking to obtain marketable products that meet promising standards. Currently, work is underway to process light and heavy reflux from catalytic cracking.

One of the directions for intensifying the process is the use of alternative raw materials, the reserves of which are inexhaustible (for example, methanol) and the creation of a thermoneutral process by combining endo- and exothermic reactions of the same purpose in one reactor.

Much attention in these studies is paid to the creation of multifunctional zeolite containing catalysts. Research in this direction has made it possible to develop a method for introducing target functions into the "memory " of the catalyst and introducing an additional function into the free vacancies of the catalyst.

BIBLIOGRAPHY

1. Aliev V.S., Indyukov N.M., Efimova S.A., Goncharov M.A., Sidorchuk I.I., Catalytic cracking in a fluidized bed. Baku - 1962., 5-11 p., 30-41 p.

2. Aliev V.S. , Rustamov M.I., Pryanikov E.I., Current state and ways of intensifying the catalytic cracking process. Azerneshr. 1966, 5-15 p.

3. Gutyrya V.S. Catalytic processes in oil refining and petrochemicals, Kyiv Naukova Dumka 1968, p.

4. Kon M.Ya., Zelkind E.M., Shershun V.G.. Oil refining petrochemical industry abroad. Moscow " Chemistry" 1968, 100-111 p.

5. ManafSüleymanov.Eşitdiklərim, oxuduqlarım, gördüklərim.Bakı-1989,64-80 s.

6. ƏjdərXəlilov.Azərbaycanınkimyaaləminəsəyahət.s.

7. Rustamov M.I. Catalytic production of high-quality motor fuels. Baku "ELM" 2006. Volume 1.

8. Rustamov M.I. Catalytic production of high-quality motor fuels. Baku "ELM" 2006. Volume 2.

9. Rustamov M.I., Aliev V.S., Guseinova A.D., Askerzade S.I., Production of AI-93 gasoline by catalytic refining of catalytic cracking gasoline on a zeolite-containing aluminosilicate catalyst. Azerbaijani oil industry. 1974. No. 5, p . 76-77.

10. Rustamov M.I., Aliev V.S., Guseinova A.D., Askerzade S.I., Production of high-octane gasoline AI-93 by catalytic upgrading of catalytic cracking gasoline on a zeolite-containing aluminosilicate catalyst. Azerbaijani oil industry. 1974. No. 7, pp . 33-34.

11.   Kardash M.M. On the issue of development of equipment and technology of oil refining in Azerbaijan. Thesis .work . 1970. p. 7-30.

12. Nigurenko V.M. History of development of research in the field of petroleum hydrocarbon production. Dis \.work . 1981.

13. Aliev V.S., Rustamov M.I. Author's certificate No. 161842.

14. Rustamov M.I., Aliev V.S., Guseinova A.D., Askerzade S.I., Abadzade Kh.I. Production of high-octane gasoline AI-93 by catalytic upgrading of catalytic cracking gasoline on an aluminosilicate catalyst. ANH. No. 6, 1972. p. 32-34

15. Chemistry and technology of fuels and oils: 1985 No. 2, 1986 No. 9, 1987 No. 7, 1988 No. 4, No. 5, 1991 No. 5, No. 8

16. Dokl. AN Az. SSR 1983 No. 4, 1984 No. 4, 1986 No. 4, 1989 No. 1.

17.   AkademikYusifMəmmədəliyev "AzərbaycanEnsiklopediyası", Nəşriyyat-Poliqrafiyabirliyi. Bakı-1996, səh 52-53.

18.   Ashurbeyli Sarah. History of the city of Baku. The Middle Ages. Azerbaijan State Publishing and Printing Association. Baku-1992. c.104-109 and c . 224-232.

19.   Aliev V.S., Pryanikov E.I., Aliev Z.E., Kabanova M.S., Indyukov N.M. Intensification of catalytic cracking of vacuum stripping. In the book: "Ways to intensify the main processes of the oil refining industry in connection with the prospects for its development." M.: TsNNITZneftegaz. 1964. p. 240-253.

20. Rustamov M.I., Aliev V.S., Guseinova A.D., Askerzade S.I., Abadzade H.Y. Production of AI-93 gasoline by catalytic upgrading of

catalytic cracking on a zeolite-containing aluminosilicate catalyst. Azerbaijani oil industry. 1974. No. 5, p. 36-37.

21. Rustamov M.I. , Aliev V.S., Guseinova A.D., Askerzade S.I., Abadzade H.Y. Production of high-octane gasoline AI -93 by catalytic upgrading of catalytic cracking on a zeolite-containing aluminosilicate catalyst. Azerbaijani oil industry. 1974. No. 7, p. 32-34.

22. Flagship of petrochemical science. Baku - "Elm" - 1999.p. 250-279

23. Problems of oil refining and petrochemistry in the works of academician M.I. Rustamov . Publishing and printing association "Azerbaijan Encyclopedia" Baku-2000.p.61-83

yes
I want morebooks!

Buy your books fast and straightforward online - at one of world's fastest growing online book stores! Environmentally sound due to Print-on-Demand technologies.

Buy your books online at
www.morebooks.shop

Kaufen Sie Ihre Bücher schnell und unkompliziert online – auf einer der am schnellsten wachsenden Buchhandelsplattformen weltweit! Dank Print-On-Demand umwelt- und ressourcenschonend produziert.

Bücher schneller online kaufen
www.morebooks.shop

Printed by Books on Demand GmbH, Norderstedt / Germany